Engineering Management in a Global Environment

Guidelines and Procedures

T0295801

Engineering Management
in a Global Environment
Guidelines and Procedures

By M. Kemal Atesmen

CRC Press
Taylor & Francis Group
Boca Raton London New York

CRC Press is an imprint of the
Taylor & Francis Group, an **informa** business

CRC Press
Taylor & Francis Group
6000 Broken Sound Parkway NW, Suite 300
Boca Raton, FL 33487-2742

Printed on acid-free paper
Version Date: 20161205

International Standard Book Number-13: 978-1-138-03574-4 (Paperback)

Library of Congress Cataloging-in-Publication Data

Names: Atesmen, M. Kemal, author.
Title: Engineering management in a global environment : guidelines and procedures / M. Kemal Atesmen.
Description: Boca Raton : Taylor & Francis, a CRC title, part of the Taylor & Francis imprint, a member of the Taylor & Francis Group, the academic division of T&F Informa, plc, [2017] Includes index.
Identifiers: LCCN 2016040405 | ISBN 9781138035744 (pbk. : alk. paper) | ISBN 978131526131 (ebook)
Subjects: LCSH: Engineering--Management. | International business enterprises--Management.
Classification: LCC TA190 .A84 2017 | DDC 620.0068--dc23
LC record available at http://lccn.loc.gov/2016040405

Visit the Taylor & Francis Web site at
http://www.taylorandfrancis.com

and the CRC Press Web site at
http://www.crcpress.com

Printed and bound in Great Britain by
TJ International Ltd, Padstow, Cornwall

Contents

Preface

The guidelines presented in this book are the result of my 33 years of experience in engineering and project management in the global arena. Challenges that I encountered through the years and the rapidly changing global business environment have made me modify and improve my engineering management procedures, which have always focused on providing constant in-depth training to bring out my subordinates' full potential. Effective engineering management in a global environment begins with an engineering organization that has been well thought out and accommodates the requirements of the company, but it also has to be flexible enough to respond to future demands in a continuously changing world.

Because of today's high-speed global business environment, engineering organizations have had to evolve from traditional technology departments to complex, international, interactive, and cross-functional matrix organizations. This book discusses the changing engineering environment and provides examples of typical and evolving engineering organizations.

After an engineering manager has put the ideal engineering organization on paper and it has been approved by upper management, each spot on that organizational chart must be filled with the right personnel. This is an unending challenge for engineering managers in a global workplace environment that involves international pools of engineering and technician talent. Guidelines for finding, interviewing, and hiring the best candidates for open positions are discussed. After an engineer or a technician has been hired, that person must be assigned to a mentor. This book discusses how to assign a mentor, a mentor's responsibilities, and interactions that occur between a mentor and a mentee. When the initial training period for new subordinates is complete, the engineering manager can begin to give assignments to them. Work assignments for novice subordinates and experienced subordinates, as well as guidelines for firefighting and dealing with unexpected changes, are detailed in the text.

Because an important function of any engineering organization is conducting effective engineering meetings, this book provides guidelines for meeting preparation, meeting management, and proper etiquette for international meetings.

Engineering managers and their subordinates must keep up with advancing technology in an ever-changing world. For this reason, this text discusses how to avoid obsolescence, how to plan for training, and how to protect a company's intellectual property. One of the most important duties of an engineering manager is to conduct thorough performance reviews for subordinates, so this text also provides guidelines for preparing performance reviews, awarding salary increases and bonuses, and giving promotions. Sooner or later, most engineering managers will have to lay off or fire employees, a topic that is discussed in the text based on my experience with having to make these hard decisions.

An engineering department's employees must have the latest tools at their disposal and enjoy a stress-free environment to perform effectively and efficiently. Guidelines for providing the essential engineering tools, developing an effective engineering department environment, and ISO 9001 and ISO 14001 requirements are detailed.

Engineering managers must cover all the bases to build first-class engineering teams. This text addresses subgroup team building, international team building, multidisciplinary team building, appointing team leaders, and procedures for cross-training.

Engineering managers cannot keep their teams isolated from others within the company nor from others outside the company. The book concludes with a discussion of relationships with upper management, other departments, customers, subcontractors, and regulatory agencies.

Acknowledgments

My over 33 years of engineering management and project management experience involving the automotive, computer, data communication, and offshore oil industries were made possible by exceptional support from my wife, Zeynep, and my family members. Sometimes it was necessary for me to be away from home more than half of the year to tackle challenging project tasks. I dedicate this book to all of the project team members with whom I have had the pleasure of working over the years, who did the hard work with enthusiasm, and who kept coming back to work with me on yet another project team without any reservations.

Author

M. Kemal Atesmen completed his high school studies at Robert Academy in Istanbul, Turkey, in 1961. He earned his bachelor of science degree at Case Western Reserve University, master's degree at Stanford University, and doctorate at Colorado State University, all in mechanical engineering. He is a life member of the American Society of Mechanical Engineers. He initially pursued both academic and industrial careers and became an associate professor in mechanical engineering before dedicating his professional life to international engineering management for 33 years. Dr. Atesmen has helped many young engineers in the international arena to bridge the gap between college and professional life in the automotive, computer component, data communication, and offshore oil industries.

Dr. Atesmen holds four patents and has published 16 technical papers and six books: Global Engineering Project Management (CRC Press, 2008), Everyday Heat Transfer Problems—Sensitivities to Governing Variables (ASME Press, 2009), Understanding the World Around Through Simple Mathematics (Infinity Publishing, 2011), A Journey Through Life (Wilson Printing, 2013), Project Management Case Studies and Lessons Learned (CRC Press, 2015), and Process Control Techniques for High-Volume Production (CRC Press, 2016).

Introduction

Because of today's high-speed global business environment, engineering organizations have had to evolve. This book discusses the role of engineering managers in the rapidly changing engineering environment. Each chapter ends with a checklist summarizing the guidelines presented in that chapter.

Chapter 1 provides insights into engineering organizations by presenting six different types of organizations commonly encountered within this global engineering environment. The roles of an engineering manager can be expected to change continuously. Today's engineering departments are comprised of different types and levels of engineers and technicians from multiple countries and cultures, and none of them has one career mentality.

For an international engineering organization, finding the right people at the right time for the right position can become quite challenging. Chapter 2 focuses on finding, interviewing, and hiring candidates for a global engineering department. Guidelines are provided for matching candidates to open positions.

When a new engineer or technician reports for work, the engineering manager's first task is to teach the new subordinate how to operate in a multidisciplinary and international environment. Chapter 3 details how to begin grooming new engineers or technicians by assigning them to compatible mentors. A mentor's responsibilities, limits of interaction between a mentor and a mentee, and the department manager's responsibility for the oversight of a mentor/mentee relationship are covered.

Today's engineering managers must assign stimulating and challenging tasks in addition to the more routine ones. Work assignments for novice and experienced engineers and technicians are covered in Chapter 4. Work assignments to put out fires and unexpected changes in work assignments are also discussed.

Properly managed, effective, and decisive meetings are vital to the success of an engineering department. Chapter 5 addresses all aspects of such meetings and provides procedures for meeting preparation, meeting management, and global meetings.

Engineering managers must plan effectively to keep their people adequately trained to operate within a multidisciplinary and global environment, in addition to keeping them challenged and motivated. Chapter 6 provides guidelines for avoiding obsolescence through training, protecting the company's intellectual property, and developing annual training plans for engineers and for technicians.

Detailed performance reviews are an important function of engineering managers. Well-prepared and well-thought-out performance reviews can further develop the talent of engineering department team members. Preparing performance reviews, awarding salary increases and bonuses, and giving promotions are detailed in Chapter 7.

Every engineering department will have subordinates with different levels of talent and different levels of ability to get things done. Inevitably, some team members will become a problem by disrupting the work environment. Engineering managers should be prepared to lay off or fire an employee, and Chapter 8 provides the guidance for doing so.

An engineering department's employees must have the latest tools at their disposal and enjoy a stress-free environment to perform effectively and efficiently. Chapter 9 discusses the engineering tools that are essential for maintaining a stimulating environment, necessary quality management systems, and continuous process improvement.

Engineering department tasks are usually executed in a team environment. Chapter 10 addresses the topics of subgroups in an engineering department, international teams, multidisciplinary teams, choosing team leaders, and cross-training.

The support of upper managers, customers, subcontractors, and regulatory agencies is crucial for the success of an engineering department. Relationships with these entities must be managed appropriately using well-established procedures. Chapter 11 provides guidelines for such procedures.

1 Typical Engineering Organizations

CHANGING ENGINEERING ENVIRONMENT

In today's global business environment and its high-speed interactions, engineering organizations have evolved from being simple 10- to 15-people traditional technology departments to complex, interactive, and cross-functional matrix organizations. Many international projects in research, development, manufacturing engineering, and customer support require various disciplines of technology. These international project teams are made up of different types and levels of engineers and technicians from several different countries. You might have an engineer working with a project team on an automated assembly project installation in Malaysia for a year, and that person might report on a dotted line to a project manager in Malaysia. To enhance your customer base and support in Japan, it might be necessary to bring a couple of Japanese engineers to your U.S. base to train for 6 months to learn the complicated test procedures for future products. These Japanese engineers would report to you on a dotted line for 6 months. Or, if your team is designing a new communication chip in the United States whose several components are being designed in Munich, Germany, then you and your team members must be in continuous communication with your counterparts in Germany.

You might be designing and building offshore oil rig equipment in the United States. After that, you have to install and train Russian engineers and technicians on Sakhalin Island in Russia. Timely and high-caliber customer support is required for your company's final payments. You might be managing a group of engineers who work from home, and you might never meet face-to-face with these subordinates during their tenure in your department. Your only interface with them is through e-mails, teleconferences, and videoconferences. Further complicating this maze of engineering management issues is that many engineers and technicians are shifting away from the one-career mentality; therefore, the turnover rate of engineers and technicians is increasing as we become more and more of a global economy.

Engineering managers must deal with globally expanding and continuously changing technologies, markets, and customers. They have to know every available technical resource in their company worldwide, in their consultants' base, in their subcontractors' base, and in the customers' base. In addition, engineering managers must be very savvy with regard to their company's worldwide supportive organizations, such as the marketing talent, financial talent, legal talent, document control talent, information technology talent, and shipping and receiving talent, among others.

1

TYPES OF ENGINEERING ORGANIZATIONS

An engineering organization chart for a medium-size production company in the United States is shown in Figure 1.1. This engineering organization is for a domestic company that designs and produces high-technology products in the United States. The organization is composed of one engineering director and five engineering departments, each with its own engineering manager. These five engineering departments have a total of 29 subordinates. The departmental divisions are for administrative purposes, but the five engineering departments share continuous cross-functional responsibilities. For example, design engineers help manufacturing engineers and quality engineers during production ramp-up of a new product. Design engineers routinely work with other engineers to modify and update a new product in order to achieve a high-yielding and reliable product. Design engineers learn during this ramp-up process all of the ins and outs of manufacturing and test processes. If one of the manufacturing or test engineers goes on vacation or quits, any one of the design engineers can step in and fill the vacuum. By wearing different hats, they become more interested in and excited about their jobs. These types of cross-functional responsibilities create a vast engineering knowledge network for a company and motivate subordinates to continue to learn and improve their skills with interesting and challenging assignments. It is the responsibility of engineering department managers to orchestrate the grooming of novice engineers to excel in multidisciplinary teams. Engineering department managers have to set up training programs and performance targets for every subordinate in order to develop an outstanding multidisciplinary knowledge base and environment aligned with professional growth targets.

Subordinates groomed in such a multidisciplinary environment will be more stimulated and more challenged. This type of engineering environment also enhances their personal growth. The down side is that subordinates will become more marketable and might jump ship if they find better opportunities. It is not possible to keep your subordinates boxed in. In today's world, managers must challenge them continuously and groom them within the multidisciplinary environment of the company. Such an environment is not confined only to engineering departments; rather, it extends to sales and marketing, legal, finance, receiving and shipping, and other domestic and international departments in the company. A multidisciplinary environment will also expand outside of the company to, for example, customers, subcontractors, regulatory agencies, and suppliers.

The classical engineering organization depicted in Figure 1.1 is changing quickly to a much flatter management organization. Engineering managers are utilizing a global pool of engineering freelancers instead of traditional full-time engineers. The advantages of using engineering freelancers include lower costs, being able to make temporary assignments for a given project, and gaining the necessary expertise very quickly. A global pool of engineering freelancers allows engineering departments to hire the best candidates for the job all over the world regardless of their proximity to the main office.

The workplaces themselves are undergoing dramatic changes to stimulate their engineers. Fully equipped exercise gyms and locker rooms, group tai chi classes, in-house masseuses, resting cots, and free lunches are becoming the norm. Dress codes,

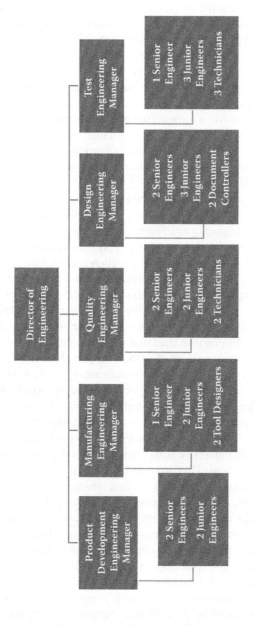

FIGURE 1.1 Engineering organization chart for a medium-size production company located in the United States.

too, are evolving; for example, male engineers are abandoning suits, ties, and dress shoes in favor of casual blue jeans, t-shirts, and flip-flops. An example of a multidisciplinary, international, flat engineering organization for a medium-size production company in the United States is shown in Figure 1.2.

Now let us analyze a different type of engineering organization where the products are becoming smaller, specifications are changing, and critical parameters are getting tighter and tighter. Such an engineering organization for a medium-size wafer production company in the United States is shown in Figure 1.3. Wafer production companies operate in a three-shift, 24/7 environment. Everything is equipment and process based. It is necessary to be a specialist in certain equipment and processes in such a production environment. You cannot ask a photolithography engineer to help a vacuum engineer when a sputtering machine is out of control. The engineering department shown in Figure 1.3 is much larger because it has to operate over three shifts every day of the week. This wafer fabrication engineering organization includes a vice president of engineering, five engineering directors, two engineering managers, and 48 subordinates.

Every subordinate must be groomed and allowed to grow within their own specialized, focused environment. As an example, a photolithography engineer in a wafer factory will specify lithography equipment, interface with the equipment vendor, and perform acceptance tests, in addition to specifying the environment (e.g., temperature and humidity control, cleanliness control, vibration control, voltage fluctuations control) for the equipment. When the equipment is accepted after several iterations, it is installed in the factory under the supervision of the photolithography engineer, who will perform process capability tests on the new equipment for critical parameters. After the process capability tests, the engineer will perform design of experiments to optimize the settings for every product that will be run through the new machine. If a photolithography process goes out of control, the photolithography engineer will be called to assist, day or night. These professionals are dedicated to their specific tasks and environment. They improve in their specialty as they gain more and more experience. They are challenged and excited by tightening critical parameters. They look to the future for improvements in lithography equipment, brought about in part by the input they provide to lithography equipment manufacturers regarding what will be required in the next six months or year. These kinds of dedicated engineers focus on their process and equipment. They strive to become leaders in their specialty.

Because these special engineers and technicians are not operating in a multidisciplinary environment, it is difficult to stimulate them and keep them at their jobs for long periods. They are hard to come by globally, which means they can easily jump ship to a competitor for higher pay or bonuses or a better work environment. Engineering managers must create a stimulating work environment for these engineers and technicians, gain their respect and trust, direct them, control them, and help them in any way possible. Greater retention of these individuals can be attained by creating a stable environment with planned changes, by having enough manpower to cover all shifts, by awarding time off after necessary all-nighters, and by helping them maintain a stable personal life by keeping a respectable distance. Engineering managers can also keep them at the leading edge of their technology by sending them to appropriate technical conferences and shows.

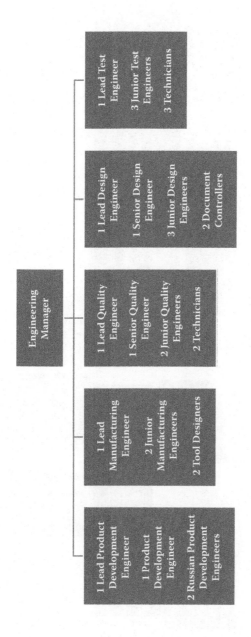

FIGURE 1.2 A multidisciplinary, global, flat engineering organization for a medium-size production company located in the United States.

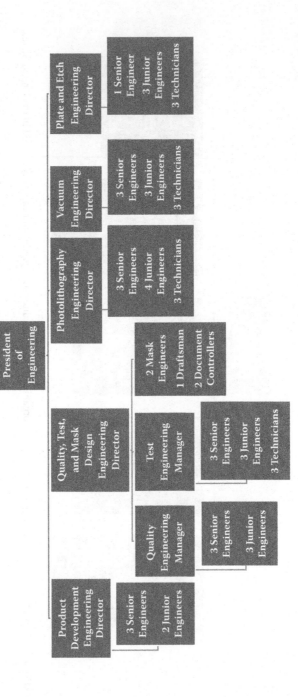

FIGURE 1.3 Engineering organization chart for a medium-size wafer production company in the United States.

It is not possible to utilize a global pool of engineering freelancers for a wafer production company such as the one shown in Figure 1.3; however, it is possible to have a multidisciplinary, flat engineering organization as shown in Figure 1.4. In an engineering organization operating in a 24/7 workplace environment, it is necessary to have proper manpower planning that allows for backup when a subordinate resigns, goes on vacation, becomes sick, goes on maternity leave, etc. Such manpower planning should strive to maintain the normal stress levels of the team; however, there will be times when the department is unavoidably short handed, and the employees must put in 70- or 80-hour work weeks. Under high-stress conditions, the leadership must be caring and provide incentives such as extra time off or bonuses to the engineers so they do not experience burnout. There is no overtime pay for engineering professionals in most companies.

We live in a global economy. It is difficult to imagine an engineering manager who does not deal daily with multiple countries and cultures. The management of international engineering teams, customers, subcontractors, and regulatory and governmental agencies is growing more sophisticated, complex, and challenging. An engineering organization for a medium-size international production company is shown in Figure 1.5. In this case, the U.S.-based company's products are developed in the United States but are produced in Malaysia and South Korea for customers in Japan. The engineering organization is structured under a U.S.-based vice president of engineering. The three engineering directors under the vice president of engineering are U.S. based, Malaysia based, and South Korea based. Two of the four engineering managers are in Malaysia and the other two are in South Korea. Under this international engineering structure, there are 11 subordinates in the United States under the product development and customer engineering director, four of whom are Malaysia based and two Japan based. There are 23 subordinates in Malaysia and 34 subordinates in South Korea.

Two Japanese engineers who support the company's Japanese customers reside in Japan and report to the engineering director in the United States. Each year, the two Japanese engineers come to the company headquarters in the United States for three months to get trained in new products and new technology. They have good command of the English language. They sometimes bring their families with them and stay in company apartments close to the U.S. facilities. The rest of the year, teleconferencing and videoconferencing allow daily communication between the Japanese engineers and the director of engineering in the United States. The Japanese engineers also visit the company's manufacturing facilities in Malaysia and in South Korea during Japanese customer visits to these production factories or if there is a quality issue with products going to Japan. They also accompany the company's engineers and managers calling on Japanese customers. In addition, they teach the company's engineers and managers about the history of Japan and business ethics in Japan, in addition to Japanese phrases that can be used during business introductions and meetings. Their yearly performance reviews are conducted in Japan by the U.S. product development engineering director. The director established their annual salary raises and bonuses to be commensurate with Japanese engineering pay and incentive levels. A simple example can demonstrate how complex and challenging it can be to manage two foreign engineers remotely and in another country. Suppose,

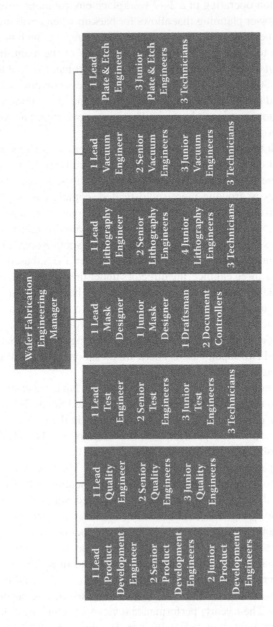

FIGURE 1.4   A flat engineering organization chart for a medium-size wafer production company in the United States.

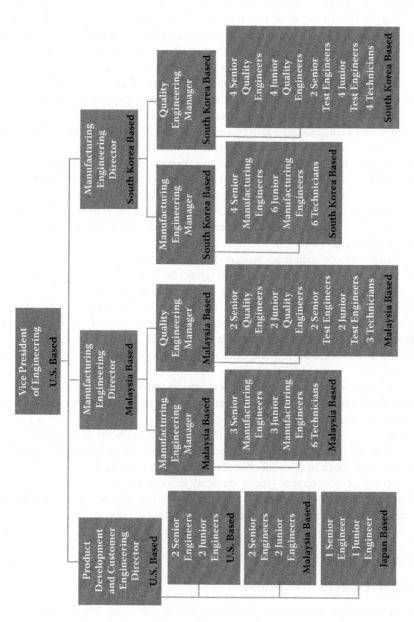

FIGURE 1.5 Engineering organization chart for a medium-size global production company.

when the engineering director is visiting Japan, he is invited to the wedding of one of his Japanese engineers. In this case, it would be to the benefit of the engineering director to learn Japanese wedding customs and the proper etiquette.

Engineering managers must be well versed in the history of countries that they are dealing with. For example, if a Japanese customer plans on visiting the South Korea plant to qualify the plant as a high-volume supplier, it is essential to have the South Korean engineers prepared for the visit. Most likely the South Korean engineers would know that Japan colonized Korea during the first half of the 20th century and established economic and military dominance there. These South Korean engineers must be advised to restrain themselves from showing any antagonistic feelings they might have toward their Japanese guests during the qualification visit.

Engineering managers must also be well versed in the religious practices and traditions of the countries they are dealing with. In Malaysia, for example, it is customary not to shake hands with a female when meeting her; it would be important to be aware of this tradition when meeting female engineers in Malaysia, although there are very few of them. Meetings with Malaysian team members must be scheduled so they do not interfere with noon prayers. Also, it is necessary to plan on and accept the fact that the efficiency of Malaysian team members will drop during the fasting month of Ramadan.

It is much more difficult to set up multidisciplinary, flat engineering organizations that operate in a global environment (see Figure 1.6). First of all, it is very inefficient to manage all of the subordinates from a U.S. base in a global engineering organization such as the one shown in Figure 1.5. It is necessary to find competent and technologically savvy engineering managers in each foreign country and to routinely interface with these global managers via teleconferencing and videoconferencing. Face-to-face visits with these managers and subordinates are necessary at least once a quarter.

Of utmost importance is training subordinates in the company's particular technical field. Multidisciplinary training will likely come in handy in global engineering environments. For our example company, it might be necessary to send U.S.-based development engineers overseas to train the Malaysian and South Korean engineers on location. Such training might include industrial statistics courses to train the Malaysian and South Korean subordinates in manufacturing quality assurance and testing. If any of the overseas groups are weak in a certain niche field, it might also become necessary to send competent consultants overseas to close the deficiency gap and bring the subordinates up to acceptable skill levels.

CHECKLIST FOR CHAPTER 1

CHANGING ENGINEERING ENVIRONMENT

- Engineering managers must deal with globally expanding and continuously changing technologies, markets, and customers.
- Engineering managers have to know every available technical resource in their company worldwide, in their consultants' base, in their subcontractors' base, and in the customers' base.

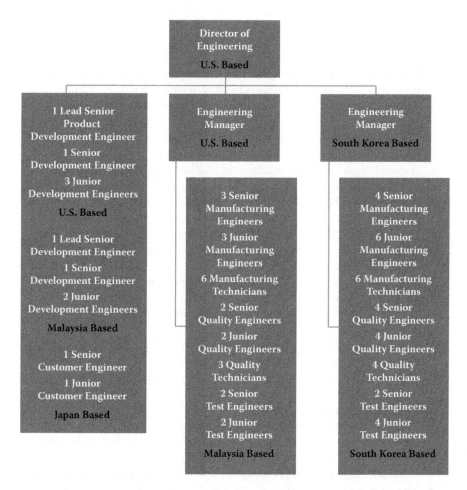

FIGURE 1.6 A flattened engineering organization chart for a medium-size global production company.

- Engineering managers must be very savvy with regard to the company's worldwide supportive organizations, such as the marketing talent, financial talent, legal talent, document control talent, information technology talent, and shipping and receiving talent, among others.

TYPES OF ENGINEERING ORGANIZATIONS

- All engineering departments should share continuous cross-functional responsibilities.
- If an engineer goes on vacation or resigns, any one of the other capable engineers in the department should be able to step in to fill the vacuum.
- By wearing different hats, engineers become more interested in and excited about their jobs.

- Cross-functional responsibilities create a vast engineering knowledge network for a company and motivate subordinates to continue to learn and improve their skills with interesting and challenging assignments.
- It is the responsibility of engineering department managers to orchestrate the grooming of novice engineers to excel in multidisciplinary teams.
- Engineering department managers have to set up training programs and performance targets for every subordinate in order to develop an outstanding multidisciplinary knowledge base and environment aligned with professional growth targets.
- A multidisciplinary environment is not confined only to engineering departments; rather, it extends to sales and marketing, legal, finance, receiving and shipping, and other domestic and international departments in the company.
- A multidisciplinary environment will also expand outside of the company to, for example, customers, subcontractors, regulatory agencies, and suppliers.
- The classical engineering organization is changing quickly to a much flatter management organization.
- Engineering managers are utilizing a global pool of engineering freelancers instead of traditional full-time engineers.
- Workplaces and dress codes are undergoing dramatic changes in order to stimulate engineers.
- Some production companies operate in a three-shift, 24/7 environment. Everything is equipment and process based. It is necessary to be a specialist in certain equipment and processes in such a production environment.
- Specialty engineers strive to become leaders in their unique and relatively small engineering environments.
- There will be times when the department is unavoidably short handed, and the employees must put in 70- or 80-hour work weeks. Under high-stress conditions, the leadership must be caring and provide incentives such as extra time off or bonuses to the engineers so they do not experience burnout.
- The management of international engineering teams, customers, subcontractors, and regulatory and governmental agencies is growing more sophisticated, complex, and challenging.
- Engineering managers, engineers, and technicians must learn about the customs and proper etiquette of the countries they are dealing with.
- Engineering managers, engineers, and technicians must be well versed in the history of the countries they are dealing with.
- Engineering managers, engineers, and technicians must be well versed in the religious practices and traditions of the countries they are dealing with.
- It is difficult to set up multidisciplinary, flat engineering organizations that operate in a global environment.
- It is very inefficient to manage all subordinates from a U.S. base within a global engineering organization.
- It is necessary to find competent and technologically savvy engineering managers in the countries that the organization is dealing with.
- Multidisciplinary training might come in handy in global engineering environments.

- It might become necessary to send U.S.-based development engineers overseas to train engineers based in a foreign country.
- If any of the overseas groups are weak in a certain niche field, it might become necessary to send competent consultants overseas to close the deficiency gap and bring the subordinates up to acceptable skill levels.

2 Hiring Engineers and Technicians for an Engineering Organization

ENGINEERING DEPARTMENT'S PERSONNEL NEEDS

The requirements of an engineering department workforce change continuously due to, for example, company expansions or contractions, offshore operations, one-time projects, the need for technical knowledge that is beyond the department's capabilities, or people jumping ship from the department. The department job openings are bounded by budget and by how upper management sees the department's personnel needs. At least once a year, the engineering manager must present to upper management the department's present and future technology base, personnel needs, and budgetary forecast. It is necessary to convince upper management that the personnel proposals are very real and required for the future of the company. In general, upper management tends to narrow down personnel requirements, asking managers to squeeze out as much work as possible from their expensive engineering personnel. In my engineering management career, I never had the luxury of an overstaffed department; I always worked with an understaffed group of engineers and technicians. On several occasions I had to deal with hiring freezes. It was easy to get an open position requisition approval for someone leaving the department, but expanding the department was another matter altogether.

The real hiring game begins after obtaining upper management job opening approvals. Managers must coordinate with human resources to initiate a talent search through the company's website, internal prospects, headhunters, company job fairs, university job fairs, conferences, trade shows, friends, subordinates' friends, and professional networks. A typical job opening announcement on a company website is shown in Figure 2.1. The engineering manager must always approve the job description before human resources posts it on the company's website. In many cases, human resources personnel do not understand what kind of talent is needed and wanted in the engineering group. An experienced headhunter can find a candidate better fit for the engineering opening in a much shorter time.

It is not unusual to receive resumés from hundreds of applicants. Human resources should be allowed to sort through the mountain of applications initially, and they should be given some definite guidelines for reviewing the job applications so good applicants are not passed over while unqualified applicants move

Position: Electric Vehicle Systems Engineering Lead

Key Qualifications:
- Five years experience with electric vehicle systems integration
- Strong aptitude in detailed and meticulous control theory with emphasis on simple and elegant design concepts for mass production
- Must be able to lead and mentor three junior engineers
- Cross-disciplinary technical skills (mechanical, materials, electrical, software)
- Excellent leadership skills with a lead-by-example attitude
- Skilled in data presentation, communication, problem solving, strategic planning
- Ability to develop collaborative networks among peers and cross-functional teams quickly
- Excellent written and verbal English language skills
- Self-starter with excellent time management skills and entrepreneurial spirit
- Relies on data to justify technical decisions
- Well versed in industrial statistics and design of experiments

Job Description:
Responsible for establishing and meeting vehicle energy control technical requirements through realization of robust designs qualified for mass production. Key responsibilities:
• Monitor mechanical and electrical designs for correct integration of vehicle systems.
• Review integration issues and drives engineering and manufacturing teams for robust closure of issues. • Drive for robust manufacturing processes. • Explore new architectures with product teams toward optimal vehicle energy usage. • Drive technical issues toward risk-free closure.

Minimum Education:
BS or MS level mechanical or electrical engineering degree required

Travel Requirements:
International travel to suppliers in Japan and Malaysia and manufacturing facilities in China as required

FIGURE 2.1   Typical website announcement for an engineering job opening.

forward in the process. Human resources will check the references and background information provided by the applicants. Most companies require drug testing for potential candidates.

In the global workplace environment, engineering managers are utilizing more and more a global pool of engineering freelancers instead of traditional full-time engineers. Advantages of using engineering freelancers for temporary assignments include lower costs and being able to gain the necessary expertise quickly. The global pool of engineering freelancers allows an engineering department to hire the best candidate for a job anywhere in the world, regardless of proximity to the office. Other ways to cut personnel costs include hiring temporary workers, consultants, interns, foreign students with appropriate work and travel visas, etc. However, the most effective way to expand an engineering department is to hire and train young college graduates who have a solid educational background, letting them develop and grow within the company's culture.

INTERVIEWING CANDIDATES

Candidate interviews can be conducted by telephone, teleconferencing, or face to face, which is probably the most effective approach. A representative from human resources should always be present during interviews. A human resources representative can answer questions about, for example, the company's health benefits, 401(k) plans, maternity leave, or vacation. Most importantly, though, the human resources representative can ensure that inappropriate and illegal questions are not asked during an interview. The author does not believe in group interviews, where several people from the department interview a candidate. It can be very overwhelming for the candidate. Even the most poised and professional job candidates can perform poorly in a group interview.

The interview should last between a half-hour and an hour. When a promising candidate has been identified, the engineering manager and several subordinates could take the candidate out to lunch or dinner. Such an extended interaction can solidify the candidate's understanding of the department's team culture, career development, work–life balance, etc. During an interview, try to avoid asking generic questions, such as "Why should I hire you?" Better yet is to probe the interviewee's accomplishments, failures, and conflicts. If the candidate had a conflict with a manager, how was it resolved? Asking in-depth questions about courses that the candidate excelled in can be very useful; for example, if a mechanical engineering graduate got a high grade in a dynamics class, initiate an in-depth discussion on the coefficient of friction to reveal the candidate's depth of technical knowledge and approach to decision making.

An interesting question to ask is "How would you move from your current home to a new home with minimal help?" The answer will demonstrate the candidate's ability to plan and get things done. The degree of energy and aggressiveness applied to the moving plan can reveal how well the candidate might keep things moving efficiently until a job is completed.

It is also important to understand an interviewee's thought processes and logic, important elements in a team environment. Organized and disciplined thought processes and logic increase an individual's productivity in a team environment and help to enhance a team's collective intelligence. As an example, a simple calculus question can proceed as follows:

1. Write on the board:

$$X = 1, \text{ then } X^2 = X$$

2. Subtract 1 from each side:

$$X^2 - 1 = X - 1$$

3. Rewrite the above equation as

$$(X - 1)(X + 1) = X - 1$$

4. Divide both sides by $(X - 1)$:

$$X + 1 = 1$$

5. Because $X = 1$, we obtain $2 = 1$!

What went wrong in this logical process?

Another classic question is as follows: Two adults and two children want to cross a river using a small dinghy. The dinghy can carry two children or one adult at a time. Everyone is a good rower. In how many trips can all of them cross the river using this dinghy? Ask the interviewee to think aloud while solving the problem. The dialog will indicate the candidate's resourcefulness, ingenuity, and ability to find ways to obtain the desired result.

HIRING A POTENTIAL CANDIDATE

After completion of a candidate's interview meetings, the engineering manager should get together with all of the interviewers to decide together whether or not to hire a particular candidate. Sometimes it can be difficult to choose from among several good candidates. Human resources will make the job offer to the favorite candidate. The job offer should reflect the department's subordinate rankings. For example, if the department's engineering level 2 requires 3 to 6 years of engineering industry experience after graduation from university, then an engineering level 2 offer should not be extended to a candidate who has only a year of engineering industry experience after completing school. No matter how much a manager may want to hire a well-qualified candidate, that candidate should not be offered an unfittingly high-ranking job within the department. Doing so will upset the subordinates' ranking and salary balance, which will cause job dissatisfaction, hamper productivity, and create jealousy among subordinates. Rather than offering a candidate an inappropriately high rank within the department as a hiring inducement, instead offer an attractive signing bonus and try to hire the person at the appropriate engineering ranking level. Other hiring perks could include a low-interest house downpayment loan or a discounted company stock purchase plan. Before making an offer to a candidate, it is important to make sure, with the help of the human resources department, that the department's salary ranges are competitive within the industry and the region.

Before hiring a prospective employee, the human resources department and legal department (intellectual property attorney) must determine whether the candidate has signed any non-disclosure or non-compete agreements with his or her previous employers that might conflict with the company's products, patents, or technology. These kinds of confidentiality agreements might affect and restrict the performance of a new employee. Non-disclosure agreements have to be clearly understood, especially between competing companies, to prevent legal problems in the future. The new candidate should also sign a non-disclosure and/or a non-compete agreement with your company in order to protect the company's technology, products, and patents.

After a candidate accepts an offer and begins work, the responsibilities of the engineering manager increase exponentially: Put together an orientation plan, along with the human resources department, for the new employee. Prepare a life-long learning proposal. Identify a mentor, who usually will be a seasoned subordinate within the department. Develop a detailed understanding of the new employee's work–life balance. Help the new employee build an ownership mentality for the department and for the company. Groom the new employee to produce excellent results in a timely fashion.

CHECKLIST FOR CHAPTER 2

ENGINEERING DEPARTMENT'S PERSONNEL NEEDS

- Engineering managers should present to upper management at least once a year their department's current and future technology base, personnel needs, and budgetary forecast.
- Engineering managers should be prepared to present open job requisitions to upper management.
- Engineering managers should obtain upper management's approvals for all departmental job openings.
- Engineering managers should approve all job descriptions before human resources posts job openings.
- Engineering managers should coordinate their department's job openings with human resources and help spread the word through all available job searching channels.
- Engineering managers should work with headhunters to fulfill urgent personnel needs.
- Engineering managers should give some definite guidelines to human resources for reviewing the job applications so good applicants are not passed over while unqualified applicants move forward in the process.
- Human resources personnel should check the references and background information provided by potential candidates.
- Engineering managers should utilize the global pool of engineering freelancers for short-term projects.

INTERVIEWING CANDIDATES

- The most effective interviews are face to face.
- It is important to always have a representative from human resources present when interviewing a job candidate.
- One-on-one job interviews can be preferable to group job interviews.
- An interview process should last between a half-hour and an hour.
- When a promising candidate has been identified, the engineering manager and several subordinates could take the candidate out to lunch or dinner.
- During an interview, it is best to avoid asking such generic questions as "Why should I hire you?"

- The candidate's answers to questions such as "How would you move from your current home to a new home with minimal help?" can demonstrate that candidate's ability to plan and get things done. The degree of energy and aggressiveness applied to solving the problem can reveal how well the candidate can keep things moving efficiently until a job is completed.
- It is also important to understand an interviewee's thought processes and logic, important elements in a team environment. Organized and disciplined thought processes and logic increase an individual's productivity in a team environment and help to enhance a team's collective intelligence.
- The interviewee should be encouraged to think aloud while solving a problem. The dialog will indicate the candidate's resourcefulness, ingenuity, and ability to find ways to obtain to the desired result.

Hiring a Potential Candidate

- After completion of a candidate's interview meetings, the engineering manager should get together with all of the interviewers to decide together whether or not to hire a particular candidate.
- Before making an offer to a candidate, it is important to make sure, with the help of the human resources department, that the department's salary ranges are competitive within the industry and the region.
- Human resources should make the job offer to the favorite candidate.
- The job offer should reflect the department's subordinate rankings.
- The candidate should not be offered an unfittingly high-ranking job within the department. Doing so will upset the subordinates' ranking and salary balance, which will cause job dissatisfaction, hamper productivity, and create jealousy among subordinates.
- Rather than offering a candidate an inappropriately high rank within the department as a hiring inducement, instead offer an attractive signing bonus or other inducement and try to hire the person at the appropriate engineering ranking level.
- Before hiring a prospective employee, the human resources department and legal department (intellectual property attorney) must determine whether the candidate has signed any non-disclosure or non-compete agreements with his or her previous employers that might conflict with the company's products, patents, or technology.
- The new candidate should also sign a non-disclosure and/or a non-compete agreement with your company in order to protect the company's technology, products, and patents.
- After a candidate accepts an offer and begins work, the engineering manager must put together an orientation plan, along with the human resources department, for the new employee.
- The engineering manager must prepare a life-long learning proposal for the new employee.
- The engineering manager must identify a mentor, who usually will be a seasoned subordinate within the department.

- The engineering manager must develop a detailed understanding of the new employee's work–life balance.
- The engineering manager must help the new employee to build an ownership mentality for the department and the company.
- The engineering manager must groom the new employee to produce excellent results in a timely fashion for the department.

3 Mentoring Engineers and Technicians in an Engineering Organization

ASSIGNING A MENTOR

One of the most important tasks of engineering managers is to groom novice engineers and technicians; however, it is almost impossible for managers to do so full time. Therefore, engineering managers must assign mentors to novice engineers or technicians. Mentoring is not every engineer's cup of tea. Mentors must be chosen carefully and must be willing to perform their mentoring duties. Mentors guide novice personnel through best practices in the engineering department and instill in them the importance of maintaining the highest technical and professional standards at all times.

Many engineers perform their tasks independently and are not always comfortable in social situations, so they may not make the best mentors. However, if a department has technology groups, a novice engineer or technician can be assigned to the leader of the appropriate technology group, although mentorship duties must be discussed with and accepted by a potential mentor before the assignment is given. For example, if a department has a finite-element stress analysis subgroup and a novice engineer is to be groomed in finite-element stress analysis technology, then that engineer could be assigned to the leader of that subgroup for mentoring. A mentor could also be a senior engineer from that subgroup. Mentorship should be assigned to individuals in the department who have demonstrated leadership skills and are willing to expand their duties. Mentors help new employees build an ownership mentality for the department and for the company.

In some cases, the mentor and mentee are not a good match. Engineering managers who observe any incompatibility between mentors and their mentees should step in and make a change. A novice engineer who is more current in his or her special field of technology might not be willing to listen to the mentor's perceived archaic advice and suggestions. Or, a high-energy mentee who tends to make quick decisions could become frustrated by a mentor who moves at a slower pace. Instead of prolonging a nonfunctioning relationship, managers should switch mentors until a proper fit is attained.

MENTOR RESPONSIBILITIES

A good mentor is patient and a good listener. Listening carefully and considering all of the issues before making a recommendation can be very powerful virtues in life. A good mentor is able to pick up on the novice engineer's or technician's technical background, technical strengths, and technical weaknesses. A good mentor creates and maintains a professional relationship with the mentee that is based on mutual respect and trust. A good mentor considers all aspects of a problem the mentee might have and provides the best possible advice for solving it. A good mentor promotes practical solutions to complicated issues. A good mentor maintains an open-door policy with the mentee, who should feel free to seek help whenever it is needed. If an issue is above and beyond the mentor's responsibility or jurisdiction (e.g., salaries, bonuses, vacation, maternity leave, issues with other departments), a good mentor will direct the mentee to the proper people (e.g., engineering department manager) for advice.

Mentors must demonstrate an ownership mentality for the department and for the company that can be passed to novice engineers or technicians. A mentor who holds grudges against upper management or who demonstrates dissatisfaction with company policies will poison the ownership mentality of the novice engineer or technician.

Mentors can give tips to their mentees about how to approach a technical problem, but mentors should not serve as instructors. Novice engineers or technicians who need further training in a certain area should take classes or attend seminars. That said, there really is no defined line between mentoring and training a mentee in engineering. Mentors can guide their mentees through complicated and special calculations, and they can go over the various procedures related to critical document control, customer contact, vendor contact, subcontractor contact, etc. However, specialized instructors should provide in-depth training in subjects requiring days or weeks of study so the mentor does not become overwhelmed trying to provide such training.

Mentors should provide details about the company's history, products, customers, and competitors; discuss the company's confidentiality procedures; and describe the company's patent application procedures. They should help mentees prepare presentations and refine their presentations through dry runs. They should also step in when mentees are running behind in an assignment to help them get back on schedule.

Mentors can be of assistance when a mentee is late to work or calls in sick. When a mentee needs additional resources (e.g., a new computer), the mentor should show the mentee how to go about requisitioning such resources. When a mentee wants to attend a technical show or a conference, the mentor should show the mentee how to obtain approval to do so. If a mentee wants to further his or her education by taking night and weekend classes and wants the company to foot the bill, the mentor should show the mentee how to approach such an endeavor. When a mentee has a conflict with a colleague, the mentor should demonstrate how to resolve such conflicts. There are numerous ways in which a mentor can help a mentee. The mentor serves as an advisor to the mentee in all company- and department-related issues.

Mentors should keep a safe distance away from their mentees' personal issues. Mentors who become too involved in their mentees' lives could open up a Pandora's box of problems. In most cases, it is advisable to direct a mentee to the human resources department when help is needed with personal issues.

Mentoring a subordinate in other locations, foreign or domestic, can be complicated. With today's fast Internet connections, it is possible to mentor a subordinate anywhere in the world through videoconferencing, teleconferencing, e-mails, and texting (keeping in mind the time differences involved). However, nothing beats face-to-face mentoring, which is much more effective and enhances the psychological camaraderie between mentor and mentee. The best possible option is to bring foreign or remotely located domestic mentees to the engineering department's facility (at most a year) for mentoring and training, so they can observe how the department thinks, how it operates, and how it solves complicated problems. The relationship between a mentor and mentee in an engineering department should last about a year, or at most two years; however, the mutual respect, trust, and friendship gained can last a lifetime. The author has observed mentors and a mentees leaving their company to form their own venture.

TYPICAL INTERACTIONS BETWEEN A MENTOR AND A MENTEE

Interactions between a mentor and a mentee can be categorized into three areas. The first category is an urgent interaction that arises when a mentee needs advice immediately. The first option for the mentee is to go to the mentor's office to discuss the urgent issue, because the mentor maintains an open-door policy for the mentee. The second option is to make a telephone call. If the mentor is busy, the mentor should direct the mentee to another source who can help. If the mentor is out of reach, then the mentee should directly go to the engineering department's manager for help.

The second category is a daily 10- to 15-minute chat in the mentee's office, most preferably during the late afternoon just before quitting time. Let's consider a mentee, Charlie, and his mentor, Don.

"How was your day, Charlie? Are your assignments keeping you on your toes?"

"I had a productive day, Don, but I was spinning my wheels for awhile to find the thermal properties of the high-strength steels I am using in my design. I searched the Internet but couldn't find a reliable source."

"Hey, I know a good source for high-strength steel properties. Roger, the senior structural designer, got a recent materials handbook that covers all of the properties of high-strength steel. See him tomorrow morning and ask to borrow the handbook."

"Thanks, Don! I had another issue with my bronze coating subcontractor. I cannot get a response from my engineering contact there regarding the status of my components. I keep calling him, and I send him e-mails. I've gotten no response for a week, and I'm worried that my components will be delayed."

"Charlie, I know the manager of the bronze coating subcontractor. Let's call him together first thing in the morning from my office. That would be 8:00 our time and 11:00 at the bronze subcontractor. Let's see if we can get the status of your components."

"Sounds good. I'll meet you in your office tomorrow morning at 8:00. I have a personal issue to discuss, too. My wife will be out of town for the next couple of weeks and I will have to pick my son up from school at 3:00. Would it be okay if I take off from work between 2:30 and 3:30 every day for the next two weeks?"

"Charlie, you should discuss this with the department manager. Tell him that you will make up for the lost time by working overtime. Emphasize to him that your tasks will not suffer and your tasks' completion dates will not change. He is a good and very understanding man. I am sure your request will not be an issue."

These daily sessions toward the end of the working day can help relieve work stress and also show a mentee that the mentor cares about him or her.

The third type of interaction is a weekly 1-hour interaction away from the working environment. It can be at the company cafeteria over a cup of coffee or a regularly scheduled lunch date each week. This type of interaction works best when it takes place toward the end of the work week, so the week's events—both positive and negative—can be discussed.

These weekly interactions should address the mentee's recent job-related successes and difficulties. The mentor should probe into whether or not the mentee likes the job. Is the mentee challenged? Are the mentee's future training and growth plans on a positive track? Does the mentee like and respect his or her colleagues, the department manager, the company management, and the company? The mentor can discuss company stock prices, the company's future, and the company's competitors, in addition to delving into the mentee's work–life balance. It is not unusual for engineers to work late into the night and on weekends; sometimes they do not even go home, instead taking a quick catnap in the office. Mentors must help their mentees achieve a balance between their work life and non-work life; otherwise, the mentee could throw in the towel and resign. The weekly meetings can also focus on lighter subjects, such as the mentee's favorite sports, hobbies, upcoming vacation, or children.

In addition to these regular interactions, mentors should be willing to attend significant events in their mentees' lives, such as birthday or work anniversary celebrations, patent plaque presentation ceremonies, and technical presentations, in addition to, when appropriate, weddings, baby showers, or funerals.

CHECKLIST FOR CHAPTER 3

ASSIGNING A MENTOR

- Mentors must be chosen carefully and must be willing to perform their mentoring duties.
- Mentorship should be assigned to individuals in the department who have demonstrated leadership skills and are willing to expand their duties.
- Mentors should help new employees build an ownership mentality for the department and for the company.
- Engineering managers who observe any incompatibility between mentors and their mentees should step in and make a change; instead of prolonging a nonfunctioning relationship, managers should switch mentors until a proper fit is attained.

MENTOR RESPONSIBILITIES

- A good mentor is patient and a good listener.
- A good mentor is able to pick up on the novice engineer's or technician's technical background, technical strengths, and technical weaknesses.
- A good mentor creates and maintains a professional relationship with the mentee that is based on mutual respect and trust.

- A good mentor considers all aspects of a problem the mentee might have and provides the best possible advice for solving it.
- A good mentor promotes practical solutions to complicated issues.
- A good mentor maintains an open-door policy with the mentee, who should feel free to seek help whenever it is needed.
- If an issue is above and beyond the mentor's responsibility or jurisdiction (e.g., salaries, bonuses, vacation, maternity leave, issues with other departments), a good mentor will direct the mentee to the proper people (e.g., engineering department manager) for advice.
- Good mentors demonstrate an ownership mentality for the department and for the company that can be passed to novice engineers or technicians.
- Specialized instructors should provide in-depth training in subjects requiring days or weeks of study so the mentor does not become overwhelmed trying to provide such training.
- Good mentors serve as advisors to mentees in all company- and department-related issues.
- Good mentors should keep a safe distance away from their mentees' personal issues; in most cases, it is advisable to direct a mentee to the human resources department when help is needed with personal issues.
- Mentoring a subordinate in other locations, foreign or domestic, can be complicated, but nothing beats face-to-face mentoring.
- The relationship between a mentor and mentee in an engineering department should last about a year, or at most two years.

Typical Interactions Between a Mentor and a Mentee

- If an urgent interaction between a mentor and mentee is necessary, the mentee should go to the mentor's office to discuss the urgent issue, because the mentor maintains an open-door policy for the mentee.
- There should be a daily 10- to 15-minute chat in the mentee's office, most preferably during the late afternoon just before quitting time.
- There should be a weekly 1-hour interaction away from the working environment. It can be at the company cafeteria over a cup of coffee or a regularly scheduled lunch date each week. This type of interaction works best when it takes place toward the end of the work week, so the week's events—both positive and negative—can be discussed.
- During the weekly interaction, the mentor should probe into whether or not the mentee likes the job and is challenged by the job, in addition to determining if the mentee's future training and growth plans are on a positive track.
- During the weekly interaction, the mentor should determine whether or not the mentee likes and respects his or her colleagues, the department manager, the company management, and the company.
- Mentors must help their mentees achieve a balance between their work life and non-work life.
- Good mentors are willing to attend significant events in their mentees' lives.

4 Work Assignments for Engineers and Technicians

Work assignments for engineers and technicians are an art. Every engineering manager learns to give effective work assignments through trial and error and experience; however, there are several simple guidelines that can be followed by every engineering manager. The following work assignment guidelines resulted from the author's 36 years of experience in engineering and engineering project management.

WORK ASSIGNMENTS FOR NOVICE SUBORDINATES

Work assignments for novice engineers and technicians should begin with easier tasks and then progress to more difficult ones. Typical early assignments might include purchasing a standard piece of equipment and following up on the purchasing process, performing simple engineering calculations and writing a report on them, or investigating an out-of-control process with production personnel, fixing the problem, and then making a short presentation about what was done at the department meeting. These are all rather simple, straightforward assignments that involve various departments of the company, ranging from sales and marketing to receiving and inspection. More complex tasks that require interaction outside of the company can then slowly be assigned; such tasks might involve suppliers, consultants, regulatory agencies, and customers. It is not a good idea to give a novice subordinate a foreign country assignment unless he or she is accompanied by a seasoned employee. When tasks are assigned to novice subordinates, committed task completion dates should be established. If it is not possible to assign a firm completion date right away, then the manager could say, for example, "I would like this new equipment to be in our plant in 3 weeks. Please investigate this equipment's purchase in detail and give me a firm delivery date in our next one-on-one meeting."

It is essential for engineering managers to have at least weekly one-on-one meetings with each novice subordinate. These one-on-one meetings should be on the calendars of both the manager and the subordinate. If for some reason the manager cannot make a weekly one-on-one meeting, it should be rescheduled without delay at the earliest possible time. In addition to these one-on-one weekly meetings, engineering managers must have an open-door policy. They should advise novice subordinates to come see them or their mentors if they run into any road blocks while carrying out their assignments. It is also advisable for engineering managers to pop into subordinates' cubicles unexpectedly for a friendly and encouraging chat once or twice a week.

During the weekly one-on-one meetings, engineering managers should discuss in detail every assignment their subordinates have, give them any necessary guidance for how to go about completing the assignments, advise them on easier ways to approach a difficult problem and how to resolve a conflict, review their reports in detail and make suggestions for improvement, review any upcoming presentations and make suggestions for improvement, and review their engineering notebooks. It is also advisable for engineering managers to help their novice subordinates prioritize their assignments. If, for example, the subordinate is modeling an engineering problem, the manager should advise them to review their modeling assumptions with a seasoned engineer in that particular field.

The supervisor and mentor of a novice engineer or technician should groom him or her to get tasks done in a timely fashion and in the right way. A reflection of successful engineering managers is if novice subordinates can demonstrate, within about 6 months after starting work, the ability to complete their assignments smoothly and with tenacity. With the engineering manager's help, novice subordinates can enhance their resourcefulness, learn the company's intricate procedures and guidelines, and identify various ways to accomplish their tasks on time.

A novice engineer or technician who does not show the ability to carry out easy projects within 6 months is not a good candidate for the department. An alternative to a probationary period for new employees is to hire interns during their last years in college in order to test their abilities and resourcefulness and to determine whether they should be kept on after they graduate.

After the 6-month trial period, successful novice subordinates can be assigned more complex and difficult tasks ranging from assignments in foreign countries to modeling intricate product components to making customer presentations. It is advisable to team novices up with seasoned personnel for these more complex assignments. The preparation, execution, and follow-up necessary for more difficult assignments require greater maturity and insight from the subordinate. This second phase of work assignments can easily take a year to master, after which the engineers or technicians should have the ability to take on any assignments with varying levels of difficulty.

WORK ASSIGNMENTS FOR EXPERIENCED SUBORDINATES

Experienced subordinates are expected to take on more challenging and difficult assignments, ranging from complicated technical designs to offshore training to customer presentations. They should be able to break down their projects into smaller tasks (at most a couple weeks in duration) and to pace themselves to complete their projects on time while maintaining the highest standards.

Even the most experienced engineers and technicians who persevere and can be counted on to complete difficult assignments may have some quirks that make them ill-suited for some tasks; for example, a particularly brilliant engineer might be a dreadful public speaker or presenter. A valuable contract could be lost by insisting that this engineer make a presentation to a potential customer. Some engineers do not have the self-confidence or patience required to prepare, practice, and make purposeful and impressive presentations. Engineering managers should not push introverted engineers or technicians into making presentations to customers or upper management or at technical conferences.

Something else to consider when assigning experienced subordinates a complicated project is that some tend to make undoable commitments. They might be great starters but poor finishers. This type of engineer or technician needs constant hand-holding and guidance by the engineering manager, who must meet with them at least once a week to stay abreast of their progress and keep them on track. At times it might be necessary to allocate other resources to these experienced subordinates if they are falling behind in critical subtasks. The engineering manager must at least mentally allocate around 25% extra time for the completion of these projects. Unfortunately, some of these types of experienced engineers hate to be micromanaged, so engineering managers must reason with them and explain that they are trying to help the subordinate complete the project successfully and on time. The significance of on-time delivery of assignments cannot be overemphasized.

Other subordinates might not be able to put together and write dependable technical reports. Technical reports are the backbone of a dependable engineering department. Technical reports are the basis for any project. They go to customers and to regulatory agencies, and they are used in patent applications and technical papers. A technical report that is full of errors, has several omissions, or is very difficult to follow and understand can endanger the good reputation of an engineering department. Some large companies have technical writers to help with such reports. During the author's career, all technical reports generated in his department were reviewed by another independent engineer. Also, all engineering reports were document controlled with the originator's signature and date, the reviewer's signature and date, and the author's signature and date. In some cases, it was necessary to assign another capable engineer the job of writing the technical report for complicated calculations performed by another engineer.

Some subordinates cannot take honest criticism of their technical work. They may not be able to acknowledge their errors, or they might not agree that there might be a better approach to a certain problem. This type of subordinate is best represented by seasoned engineers and technicians who do not want to be told by a junior engineer or technician (or sometimes even the engineering manager) how to approach a problem. Other subordinates might be left behind as technology advances, or they suffer from a "not invented here" syndrome. In such cases, the engineering manager must take the lead and give proper credit to other contributors. Seasoned engineers and technicians must be urged to catch up with advancing technology during their reviews.

Another type of subordinate might prefer not to go on international assignments. This reluctance for offshore assignments might be due to personal reasons. The author remembers an engineer who had never ridden on an airplane in his life and never intended to. Another engineer wanted to be with his family for each of his religious holidays, so he could not be assigned long offshore assignments.

Personal lives can have an effect on performance; for example, an experienced engineer botched an important technical presentation in front of a customer during the critical design review. Apparently, the night before the presentation he and his wife decided to divorce after a quarrel. In the morning, he gave no clue as to what had happened the night before; however, despite covering his devastation as professionally as possible, he could not control his stress level during the customer presentation. Subordinates must be mentally prepared for important events, and engineering managers should try to be aware of the psychological state of their subordinates before such events.

WORK ASSIGNMENTS FOR FIREFIGHTING

Putting out fires, or firefighting, is a fact of life in any engineering department. How the engineering manager attacks a firefighting issue depends on the depth of talent in the department, on fail-safe planning, and on the engineering manager's experience. Unforeseeable firefighting issues that pop up with customers, subcontractors, and other departments must be resolved without any delay. Not surprisingly, most stressful firefighting situations occur in short-handed engineering departments.

Having to divert a subordinate or a team of subordinates to firefight an issue can delay other projects in the department and increase the stress levels of all subordinates. If the department is too short-handed to respond adequately to a critical issue, it might become necessary to ask for additional help from other resources in the company, to seek the help of consultants, or to call back some subordinates from their time off. It might even be necessary to request help from customers.

Whatever the case, subordinates must drop whatever they are doing to address the urgent issue at hand. Assigning such urgent tasks can be tricky for engineering managers, who must find the right talent for the urgent job without upsetting the department's apple cart. Urgent tasks cannot be assigned to subordinates who are focused on completing complicated and difficult tasks that are on the critical path of a project, but they also cannot be assigned to novice subordinates who lack the necessary experience.

Sometimes putting out the fire is not a permanent solution to the problem. The same issue could keep occurring over and over again. In such cases, a capable team must be assembled to attack the problem with a detailed cause-and-effect analysis and a plan to identify the root cause of the problem. Persistent issues cannot be fixed in a day or two, and it could take a long time to find a permanent solution. Engineering managers must assess these situations carefully and allocate resources as necessary to eliminate the problem.

Engineering managers must be prepared not only to identify risky domains under their jurisdiction but also to report any risky domains they might observe under the control of other managers. It is much more difficult to identify risky domains in international operations. The ability to identify risky domains comes from experience, from listening, and from knowing what is going on in the department and in other departments, including international operations. Because it is not possible for engineering managers to keep up with every detail companywide, they must trust the judgment of other managers, especially international ones. However, engineering managers who spot a deficiency in management in other departments or in offshore operations should immediately report the situation to their superiors to get it corrected. Ignoring such things can thrust an engineering department into firefighting mode at the worst possible time.

UNEXPECTED CHANGES IN WORK ASSIGNMENTS

In an engineering department, some assignments might never be finished, and some assignments can suddenly veer off in another direction. Such shifts in assignments can be stressful to subordinates. Engineering managers must carefully maneuver their subordinates through these unexpected changes, which can arise for many reasons, including improved technology, altered specifications, newly identified customer needs, a refocus by upper management, or subcontractor restraints. Almost-completed projects can be shelved due to budget cuts by upper management. The author has experienced a communication chip design and an automobile engine design being shelved after successful prototyping due to a change in marketing direction. At times, changes in specifications and scope that come from customers, subcontractors, and regulatory agencies can force redesigns from scratch. Such modifications can affect subordinates. Some are able to take on these changes without looking back, but others become discouraged—"I worked on this project day and night to complete it and now it's been thrown out the window. What a waste!"

Engineering managers must train their subordinates to accept that a change in a given assignment is inevitable. Engineering assignments are performed in a continuously flowing, changing, and challenging environment. For example, implementing new technology right in the middle of project execution can be tricky. Perhaps the newly adopted technology has not matured enough. The unavoidable learning curve for a new technology cannot be overlooked, and budgeting or software issues can further complicate the implementation of a new technology. Engineering managers must keep their subordinates sane and proud of what they have achieved in a constantly evolving engineering world.

CHECKLIST FOR CHAPTER 4

WORK ASSIGNMENTS FOR NOVICE SUBORDINATES

- Work assignments for novice engineers and technicians should begin with easier tasks and then progress to more difficult ones.
- Easy tasks should be rather simple, straightforward assignments that involve various departments of the company, ranging from sales and marketing to receiving and inspection.
- More complex tasks that require interaction outside of the company can then slowly be assigned; such tasks might involve suppliers, consultants, regulatory agencies, and customers.
- It is not a good idea to give a novice subordinate a foreign country assignment unless he or she is accompanied by a seasoned employee.
- When tasks are assigned to novice subordinates, committed task completion dates should be established.
- Engineering managers should have at least weekly one-on-one meetings with each novice subordinate.
- These one-on-one meetings should be on the calendars of both the manager and the subordinate.

- If for some reason the manager cannot make a weekly one-on-one meeting, it should be rescheduled without delay at the earliest possible time.
- Engineering managers must have an open-door policy. They should advise novice subordinates to come see them or their mentors if they run into any road blocks while carrying out their assignments.
- Engineering managers should pop into subordinates' cubicles unexpectedly for a friendly and encouraging chat once or twice a week.
- During the weekly one-on-one meetings, engineering managers should discuss in detail every assignment their subordinates have and help them prioritize their assignments.
- If a subordinate is modeling an engineering problem, the engineering manager should advise them to review their modeling assumptions with a seasoned engineer in that particular field.
- Novice subordinates should demonstrate the ability to get easy assignments done smoothly and with tenacity.
- With the engineering manager's help, novice subordinates can enhance their resourcefulness, learn the company's intricate procedures and guidelines, and identify various ways to accomplish their tasks on time.
- A novice subordinate who demonstrates the ability and resourcefulness to carry out simple projects during the probationary period should be a good candidate to work for the department.

WORK ASSIGNMENTS FOR EXPERIENCED SUBORDINATES

- Experienced subordinates should be able to break down their projects into smaller tasks (at most a couple weeks in duration) and to pace themselves to complete their projects on time while maintaining the highest standards.
- Engineering managers should not push introverted engineers or technicians into making presentations to customers or upper management or at technical conferences.
- Some experienced subordinates tend to make undoable commitments; they might be great starters but poor finishers. This type of engineer or technician needs constant hand-holding and guidance by the engineering manager to keep them on track.
- The engineering manager must at least mentally allocate around 25% extra time for the completion of projects by experienced subordinates who make unrealistic commitments.
- Some experienced engineers and technicians hate to be micromanaged. Engineering managers must reason with them and explain that they are trying to help the subordinate complete the project successfully and on time.
- The engineering manager cannot overemphasize the significance of on-time delivery of assignments.
- Some experienced and brilliant subordinates might not be able to put together and write dependable technical reports.
- All technical reports generated in an engineering department should be reviewed by another independent engineer.

- All engineering reports should be document controlled.
- Engineering managers might have to assign another capable engineer the job of writing the technical report for complicated calculations performed by another engineer.
- Some subordinates cannot take honest criticism of their technical work.
- Some subordinates might be left behind as technology advances, or they suffer from a "not invented here" syndrome.
- The engineering manager must take the lead and give proper credit to other contributors.
- Seasoned engineers and technicians must be urged to catch up with advancing technology during their reviews.
- Engineering managers should make sure that subordinates are willing to go on international assignments, without reservation.
- Subordinates must be mentally prepared for important events, and engineering managers should try to be aware of the psychological state of their subordinates before such events.

Work Assignments for Firefighting

- Firefighting is a fact of life in any engineering department.
- Most stressful firefighting situations occur in short-handed engineering departments.
- Having to divert a subordinate or a team of subordinates to firefight an issue can delay other projects in the department and increase the stress levels of all subordinates.
- If the department is too short-handed to respond adequately to a critical issue, it might become necessary to ask for additional help from other available resources.
- The engineering manager must find the right talent for the urgent job without upsetting the department's apple cart.
- Urgent tasks cannot be assigned to subordinates who are focused on completing complicated and difficult tasks that are on the critical path of a project.
- Urgent tasks cannot be assigned to novice subordinates who lack the necessary experience.
- For some persistent problems, a capable team must be assembled to attack the problem with a detailed cause-and-effect analysis and a plan to identify the root cause of the problem.
- Persistent issues cannot be fixed in a day or two, and it could take a long time to find a permanent solution.
- Engineering managers must assess these situations carefully and allocate resources as necessary to eliminate the problem.
- Engineering managers must be prepared not only to identify risky domains under their jurisdiction but also to report any risky domains they might observe under the control of other managers.
- It is much more difficult to identify risky domains in international operations.

- The ability to identify risky domains comes from experience, from listening, and from knowing what is going on in the department and in other departments, including international operations.
- Because it is not possible for engineering managers to keep up with every detail companywide, they must trust the judgment of other managers, especially international ones.
- Engineering managers who spot a deficiency in management in other departments or in offshore operations should immediately report the situation to their superiors to get it corrected.
- Ignoring such things can thrust an engineering department into firefighting mode at the worst possible time.

UNEXPECTED CHANGES IN WORK ASSIGNMENTS

- In an engineering department, some assignments might never be finished, and some assignments can suddenly veer off in another direction.
- Such shifts in assignments can be stressful to subordinates.
- Engineering managers must carefully maneuver their subordinates through these unexpected changes.
- Sudden changes can arise for many reasons, including improved technology, altered specifications, newly identified customer needs, a refocus by upper management, or subcontractor restraints.
- Engineering managers must train their subordinates to accept that a change in a given assignment is inevitable.
- Engineering assignments are performed in a continuously flowing, changing, and challenging environment.
- Implementing new technology right in the middle of project execution can be tricky.
- Newly adopted technology may not have matured enough.
- The unavoidable learning curve for a new technology cannot be overlooked.
- Budgeting or software issues can further complicate the implementation of a new technology.
- Engineering managers must keep their subordinates sane and proud of what they have achieved in a constantly evolving engineering world.

5 Meetings

Meetings are the backbone of an engineering department. Meetings stimulate the decision-making process and hold the department together. Properly managed, effective, and decisive meetings can contribute to the success of an engineering department. In contrast, improperly managed, overly long, and indecisive meetings can demoralize an engineering department. Meetings can take up more that 80% of an engineering manager's workday. Figure 5.1 shows the schedule for a typical workday of an engineering manager. It is not unusual for meeting times to conflict due to emergencies, customer meetings running long, etc. Engineering managers must be able to juggle both local and global meetings and prioritize their presence at meetings in an efficient and artful way. They have to understand when and to whom to delegate meeting responsibilities when necessary.

MEETING PREPARATION

An engineering department's meetings come in different flavors—department meetings held locally and globally on fixed days and times of the week, design review meetings, customer meetings, subcontractor meetings, regulatory agency meetings, upper management meetings, subgroup meetings, individual one-on-one meetings, fun and morale-boosting meetings, etc. Every meeting (except emergency ones) should be scheduled and announced by the engineering manager or by subordinates using a consistent procedure and meeting announcement form. A typical meeting announcement form is shown in Figure 5.2. Meetings should be announced at least a week before they are scheduled to be held so any meeting time or conference room conflicts can be resolved in a timely fashion. Also, regular reminders for routinely scheduled meetings should be provided by the originator.

When emergency meetings are necessary, the engineering manager should visit the subordinates' work areas to invite them to the emergency meeting and tell them the time, place, and agenda. Emergency meetings should be conducted just as regular meetings are (e.g., recording meeting minutes, developing an action item list).

Before an engineering design review meeting with a customer or with upper management, it is always advisable to do a dry run. Make sure that the customer approves the meeting agenda prior to the meeting. Prepare a separate office or a conference room for the customer's participants to discuss issues in privacy among themselves. Make communication tools such as Internet connections, telephones, fax machines, and printers available to them in a private setting.

Minor things can cause unexpected delays during a meeting. Be sure to have extra bulbs on hand for a slide projector, overhead projector, or laptop computer external projection system. Know how to use all of the presentation tools; try them out before

Date: 2 June 2016 Thursday				
Time	Meeting With	Meeting Location	Meeting Media	Meeting Times in Global Locations
8–9 AM	Hans Marken *German Designer Weekly*	Office	Teleconference	Germany 17–18 PM Thursday
9–10 AM	Al Crane *Weekly One-on-One*	Office		
10–11 AM	Silvia Dante *Weekly One-on-One*	Office		
11–12 noon	John Olivers *Weekly Upper Management*	Executive Conference Room #1		
12–13 PM	Joe Upstain *Farewell Lunch*	Beachside Cafe		
13–14 PM				
14–15 PM	MX Product Design Review	Engineering Conf. Room #2		
15–16 PM	MX Product Design Review	Engineering Conf. Room #2		
16–17 PM	Malaysia Manufacturing Engineering Weekly	Videoconference Room #2	Videoconference	Malaysia 8–9 AM Friday
17–18 PM				
18–19 PM				
19–20 PM				
20–21 PM	India Software Design Development	From Home	Telephone	India 8:30–9:30 AM Friday
21–22 PM				

FIGURE 5.1   An engineering manager's meeting schedule for a normal workday.

the meeting to make sure that everything is operational. Sometimes simple things like blackboard cleaners, felt pens, notepads, pens, pencils, and erasers are over-looked and can cause disruptions and delays during a meeting.

Catering is a crucial element of important or long meetings. Coffee, tea, snacks, lunches, and sometimes dinners should be served at the meeting, and the meeting organizer should announce the availability of food at the beginning of a meeting. Alternatively, it might be desirable to take customers to a local restaurant for a long lunch break. For long meetings, breaks at appropriate intervals are a must so partici-pants can stretch, use the restroom, and re-energize.

Meeting Subject: Weekly Status for Automated Suspension Assembly System Installation in Malaysia
Time: 4–5 PM PST (Pacific Standard Time) on 2 June 2016 Wednesday *and* 8–9 AM Malaysian Time on 3 June 2016 Thursday
Place: Engineering Conference Room #1 in California and South Conference Room in Penang
Meeting Media: Videoconferencing
Meeting Originator: Alan Summer, Director of Engineering
Requested Meeting Participants: *U.S. Manuf. Engr. Dept.:* Shirley Ping, John Shonehirst *Malaysian Manuf. Engr. Dept.:* Aziz Tariq, Ahmad Bulani *U.S. Quality Engr. Dept.:* Steve Forrest, Zelda Trecsey *Malaysian Quality Engr. Dept.:* Fatima Payyo, Husnu Mantayi *U.S. Purchasing Dept.:* Dennis England
Meeting Recorder: Shirley Ping
Meeting Agenda: 4:00–4:15 (8:00–8:15) Spare parts purchasing status with delivery dates to Penang—Dennis England 4:15–4:30 (8:15–8:30) System voltage control equipment installation status—Aziz Tariq 4:30–4:50 (8:30–8:50) System qualification runs status—Fatima Payyo 4:50–5:00 (8:50–9:00) New issues and action items
Meeting Minutes:
Action Item List:
Document Control Folder: Penang—Automated Suspension Assembly System

FIGURE 5.2   Typical engineering department meeting announcement form.

MEETING MANAGEMENT

Meeting management is an art by itself. In most engineering meetings, upper management or senior personnel dominate the meeting, but novice subordinates should be encouraged to participate in discussions during a meeting. Meetings should be planned to avoid wasting the time of participants who do not have anything to do with a particular issue. Meeting minutes and action items for every regular or emergency meeting should be recorded by a meeting participant during the meeting. The meeting recorder should be announced at the beginning of the meeting. After the meeting, meeting minutes and action items should be approved by the appropriate people, such as the meeting originator, customer representative, subcontractor representative, or regulatory agency representative. The minutes should then be filed according to the company's document control procedures. Meeting minutes and action items should be distributed to all participants, to all action item owners, and to appropriate upper management. Some action item owners might not have participated in the meeting. In this case, it is the responsibility of the engineering manager or meeting organizer to discuss the action item with the owner to see if he or she can complete the action item within the allocated time frame. When an action item is given to someone who did not participate in a meeting, it is always a good idea to discuss the action item with that person after the meeting to establish a completion time. A sample engineering department's meeting minutes and action items form is presented in Figure 5.3.

Meeting Subject: Weekly Status for Automated Suspension Assembly System Installation in Malaysia
Time: 4–5 PM PST (Pacific Standard Time) on 2 June 2016 Wednesday *and* 8–9 AM Malaysian Time on 3 June 2016 Thursday
Place: Engineering Conference Room #1 in California and South Conference Room in Penang
Meeting Media: Videoconferencing
Meeting Originator: Alan Summer, Director of Engineering
Requested Meeting Participants: *U.S. Manuf. Engr. Dept.:* Shirley Ping, John Shonehirst *Malaysian Manuf. Engr. Dept.:* Aziz Tariq, Ahmad Bulani *U.S. Quality Engr. Dept.:* Steve Forrest, Zelda Trecsey *Malaysian Quality Engr. Dept.:* Fatima Payyo, Husnu Mantayi *U.S. Purchasing Dept.:* Dennis England
Meeting Recorder: Shirley Ping
Meeting Agenda: 4:00–4:15 (8:00–8:15) Spare parts purchasing status with delivery dates to Penang—Dennis England 4:15–4:30 (8:15–8:30) System voltage control equipment installation status—Aziz Tariq 4:30–4:50 (8:30–8:50) System qualification runs status—Fatima Payyo 4:50–5:00 (8:50–9:00) New issues and action items
Meeting Minutes: 4:00–4:15 (8:00–8:15) Spare parts purchasing status with delivery dates to Penang—Dennis England updated spare parts purchasing status. Fatima stressed the material certificates issue with high-strength steels. Dennis said he is on top of them. Ahmad asked Dennis to expedite heating lamps that are scheduled to be delivered by 15 June 2016. 4:15–4:30 (8:15–8:30) System voltage control equipment installation status—Aziz Tariq updated system voltage control equipment installation. All tasks are on schedule. Training on operation and maintenance will be completed by the end of the month. 4:30–4:50 (8:30–8:50) System qualification runs status—Fatima Payyo will present system qualification runs plan during next week's meeting. 4:50–5:00 (8:50–9:00) New issues and action items—Alan indicated that they are very close to hiring another manufacturing engineer for the Malaysian team for the automated suspension assembly system night shift operation.
Action Item List: 1. Expedite delivery of heating lamps. Needed in Penang by 15 June 2016—Dennis England 2. Get a separate power line control box installed by an outside electrician by 12 June 2016—Aziz Tariq 3. Get trained by the manufacturer of voltage control equipment on operation and maintenance by 30 June 2016—Aziz Tariq and Husnu Mantayi 4. Prepare a system qualification runs plan, distribute and present by 9 June 2016—Fatima Payyo 5. Alan Summer has to get the night shift manufacturing engineer on board by 16 June 2016
Document Control Folder: Penang—Automated Suspension Assembly System

FIGURE 5.3 Typical engineering department meeting minutes and action item list.

Sometimes in global meetings, simultaneous translations might be needed, so each item of discussion will take longer, and the time allocated for each discussion item must be adjusted accordingly. Critical and long meetings (e.g., specification clarification discussions with a customer or subcontractor) should be tape-recorded with the permission of the participants, so that later on there will be no controversy regarding issues decided upon during the meeting and nothing important will be missed by a perhaps overwhelmed meeting recorder.

The meeting recorder should be in attendance throughout the entire meeting. After the meeting minutes and action item list have been approved by the meeting organizer and all of the action item owners, they should be distributed to all involved parties and released to engineering document control. For customer and subcontractor meetings, meeting minutes and action item documents should also be approved by the customer or subcontractor before being distributed and released to document control. The engineering manager must follow up to make sure all action items are completed to the highest standards and on time by all action item owners.

Not everyone must stick around for an entire meeting. In the example shown in Figure 5.3, the purchasing agent, Dennis England, presented the spare parts purchasing status for the automated suspension assembly system at the beginning of the meeting. After his presentation was completed, he left the meeting and returned to work because later items on the agenda did not pertain to him.

Being late to a meeting is not acceptable. The originator of a meeting who realizes he or she will be late, even by just a couple of minutes, should ask a subordinate to begin the meeting per the released meeting agenda. Subordinates who will be late to a meeting should inform the originator beforehand. In Malaysia, the author had much difficulty in getting subordinates and people from other departments to attend meetings on time. Repeated warnings were ineffective. Threatening to fine them a dollar for every minute they were late (funds that would later be used to take everyone out to lunch) didn't work, either. To address this rather pervasive lackadaisical attitude, the general manager of the plant issued a stern memorandum about meeting attendance. His memorandum was right to the point: "Anyone who is late to a meeting that he or she is supposed to be attending will be dismissed from the company the next day." This memorandum from the general manager did the trick, and everyone was always on time to meetings after that.

Engineering department meetings can waste a lot of time. Engineering managers need to plan efficient meetings and make sure that subordinates, locally and globally, are not twiddling their thumbs. A common complaint of engineers and technicians is "Meetings, meetings, meetings! I am overwhelmed with meetings. I can't get any of my work done." It helps to break up meetings into well-defined segments so people can attend the meeting only during the particular segment that is relevant to them.

If discussion of a subject during a meeting runs over its allotted time, it is best to politely inform the people involved that another meeting will be scheduled to complete the ongoing discussion and then move on to the next agenda item. It is not unusual for upper managers to deviate from the subject under discussion and digress to other issues, so they must be gently steered back to the topic at hand.

Even for small meetings, such as a one-on-one meeting between the engineering manager and a subordinate, it is necessary to fill out the official company meeting form (see Figure 5.2) and release it to the appropriate document control folder. As mentioned earlier, one-on-one meetings with subordinates should occur at least once a week at a predetermined time to discuss and review their ongoing assignments.

Another meeting that should be routinely scheduled (at least biweekly) is the engineering department meeting, at which the engineering manager can address important departmental, company, customer, and subcontractor issues. All subordinates should be given the chance to share what they have accomplished since the last departmental meeting.

GLOBAL MEETINGS

Meeting protocols can vary quite a bit around the world. It is necessary to learn and observe the meeting ground rules of the country where the meeting is being held. For example, when attending customer design review meetings at customer sites in Japan, the project team should know to take their seats on the side of the table away from the conference room entry door when they are escorted into the conference room, and they should sit according to rank and experience. Shortly thereafter, the Japanese customer's team enters and sits across the table from the project team, again according to rank and experience. Handshakes, introductions, and card exchanges are very formal. Being able to speak several Japanese phrases during introductions is always helpful. Business cards should be dual language and include the employees' pictures on them. Names and their pronunciations are very important in Japan, as they are at any international meeting. Learning the customer's meeting participants' names and their correct pronunciations, as well as the positions held by the meeting participants, ahead of the meeting is very helpful.

After opening welcoming remarks are made by the ranking member of the customer's party, the ranking member of the engineering team is given the opportunity to make opening remarks. Everything spoken in Japanese is translated into English or *vice versa* by a bilingual member of the customer's team. When the meeting is at a customer's facility, the customer's meeting organizer leads the meeting. If additional agenda items have come up or if any changes to the existing agenda need to be made, these should be addressed at the beginning of the meeting.

Participants must be prepared for engineering design review meetings that last 10 to 12 hours, so it is best for meeting participants to not be suffering from jet lag; it is important to remain sharp throughout such long meetings. During a meeting, if a team member thinks that he or she is losing a debate or is in doubt as to what position to take on an issue, the team should regroup and brainstorm the problem. The team should call for a timeout and request to go to a private room to discuss the crucial point among themselves or seek the advice of other people in the company by telephone, text messaging, or e-mails.

In the final phase of the meeting, all action items, their assigned owners, and their completion dates should be agreed upon by everyone in attendance. Meeting minutes and action item lists should be completed when everyone is there and everything is fresh in their minds. In some cases, when the action item owner is not at the meeting, finalizing a completion date can be delayed until the action owner has been contacted. Meeting documents should be precise, accurate, and detailed. Completed and approved documents should be distributed to the appropriate people in the company and released to the appropriate document control folder. The engineering manager's job is not yet finished, though, as he or she must follow up to make sure all action items are completed to the highest standards and on time by all action item owners.

CHECKLIST FOR CHAPTER 5

MEETING PREPARATION

- Every meeting (except emergency ones) should be scheduled and announced by the engineering manager or by subordinates using a consistent procedure and meeting announcement form.
- Announce meetings at least a week before they are scheduled to be held so any meeting time or conference room conflicts can be resolved in a timely fashion.
- Before an engineering design review meeting with a customer or with upper management, it is always advisable to do a dry run.
- Make sure that the customer approves the meeting agenda prior to the meeting.
- Prepare a separate office or a conference room for the customer's participants to discuss issues in privacy among themselves. Make communication tools such as Internet connections, telephones, fax machines, and printers available to them in a private setting.
- Know how to use all of the presentation tools required for the meeting; try them out before the meeting to make sure that everything is operational.
- Be sure that all of the necessary blackboard cleaners, felt pens, notepads, pens, pencils, erasers, etc., are on hand.
- Arrange for the catering of any food and drink required for the meeting.
- Allow time for breaks during long meetings.

MEETING MANAGEMENT

- For a global meeting, the engineering manager should set meeting times that are appropriate for all of the time zones involved.
- The meeting recorder should be announced at the beginning of the meeting.
- Simultaneous translations might be needed, so each item of discussion will take longer, and the time allocated for each discussion item must be adjusted accordingly.
- Critical and long meetings should be tape-recorded with the permission of the participants.
- Even though upper management or senior personnel might be dominating the meeting, novice subordinates should be encouraged to participate in discussions during the meeting.
- For long meetings, participants may attend only that portion of the meeting that pertains to them.

- Meeting minutes and action item documents should also be approved by the customer or subcontractor before being distributed and released to document control.
- An action item owner who is not in attendance should be contacted after the meeting to establish a completion date for the action item.
- The engineering manager must follow up to make sure all action items are completed to the highest standards and on time by all action item owners.
- Because being late to a meeting is not acceptable, the engineering manager must take actions to prevent late meeting arrivals.
- Engineering departments can have a lot of meetings. Except for unusual cases, such as customer design review meetings, subordinates should not spend more than 4 hours a week in meetings.
- If discussion of a subject during a meeting runs over its allotted time, it is best to politely inform the people involved that another meeting will be scheduled to complete the ongoing discussion and then move on to the next agenda item.
- Even for small meetings, such as a one-on-one meeting between the engineering manager and a subordinate, it is necessary to fill out the official company meeting form and release it to the appropriate document control folder.

Global Meetings

- Engineering managers should learn and observe the meeting protocols and customs of the country where the meeting is being held.
- Meetings held in a foreign country should not be scheduled to take place on national or religious holidays.
- Meeting participants should not be suffering from jet lag; it is important to remain sharp throughout such long meetings.
- During a meeting, if a team member thinks that he or she is losing a debate or is in doubt as to what position to take on an issue, the team should regroup and brainstorm the problem.
- The team should call for a timeout and request to go to a private room to discuss the crucial point among themselves or seek the advice of other people in the company by telephone, text messaging, or e-mails.
- Meeting minutes and action item lists should be completed when everyone is there and everything is fresh in their minds.
- In some cases, when the action item owner is not at the meeting, finalizing a completion date can be delayed until the action owner has been contacted.
- Meeting documents should be precise, accurate, and detailed.
- The engineering manager must follow up to make sure all action items are completed to the highest standards and on time by all action item owners.

6 Keeping Up with Technology and a Changing World

AVOIDING OBSOLESCENCE

An important part of the engineering manager's job is to ensure that the department's engineers and technicians continue their education and training to remain relevant in a changing world. Obsolescence can be devastating not only for subordinates but also the company as a whole. It is important to keep learning in life's university. Each review period, engineering managers should light a fire under their subordinates by discussing, among other things, what classes they should take, which technical conferences or trade shows they should attend, what kind of papers they should strive to publish, and what possible patentable areas should be further developed.

Effective engineering managers will encourage their subordinates to maintain a separation between their work passions and their hobbies outside of work. Managers can show an interest in their subordinates' interests outside of work (e.g., oenology, tennis, music), but these non-work-related passions should not interfere with work. Whenever possible, work hours should not disturb the life rhythm of a subordinate, and, in turn, that life rhythm should not overwhelm the employee's job responsibilities. A subordinate who goes out for a run every day at lunch and takes an extra 20 minutes to shower and return to work should only be allowed to do so with the understanding that he will make up that extra time.

Engineering managers should be able to pick up on characteristics of their subordinates' personalities. A subordinate who has an outgoing character, likes people, and has a knack for organization might be groomed toward being a group leader, a project manager, or a department manager. A subordinate who prefers technical tasks, is skilled in science and mathematics, and has the ability to get complex things done can be encouraged to excel in multidisciplinary technologies pertinent to the company.

When an engineering group is developing a complex system, such as a smart grid, they must have a good grasp of current power technology, information technology, communication technology, sensors technology, etc., in order to be able to design the best smart grid possible. Everyone must know word by word all Institute of Electrical and Electronics Engineers (IEEE) standards that are related to the design of a smart grid.

If an engineering group is designing new equipment for a European Union customer but the designers are trained only in American Society of Mechanical Engineers standard for dimensioning and tolerancing (ASME Y14.5), they must familiarize themselves with the International Organization for Standardization dimensioning and tolerancing standard (ISO 8015) before the project begins. In this global engineering environment, engineers and designers must thoroughly understand the differences between the two dimensioning and tolerancing standards. Suppose an engineering group is designing offshore equipment to be installed off of Sakhalin Island, Russia. In this case, the designers must be experts in the elastic and plastic properties of the high-strength casting materials that will be used in the design, as these materials will have to perform in very cold temperatures (e.g., –50°C) and under 100-year wind, wave, and seismic loads.

PLANNING FOR TRAINING

Training a subordinate requires careful planning and preparation. Because of varying departmental job responsibilities, looming project task deadlines, and unusually stressful situations, it might be necessary to reschedule a training event for a subordinate. However, training should not normally be a secondary priority. When subordinates realize that their training has a low priority, they are likely to leave the company. Training for subordinates should extend beyond science and technology and cover world markets, environmental issues, legal issues, health issues, global customs and traditions, and, most importantly, your customers and competitors. Engineering managers should challenge their subordinates beyond their current capabilities, but at the same time subordinates should receive the support necessary to complete simple tasks on time and with outstanding results.

The design parameters for a new car being developed for emerging markets such as Brazil, Russia, India, and China (BRIC) can be quite different than for a new car destined for the U.S. market; for example, emissions and passenger safety requirements will differ. No matter the market, however, ensuring the health and safety of potential customers should be a top priority for any engineering organization.

In China, the requirements for the quality of a factory's water discharge might not be as stringent as those in the United States or in EU countries, but the team's goal should be not to endanger the environment and to maintain the highest possible environmental standards. The highest technical standards should be adhered to at all times, and under no conditions should any type of bribery be accepted.

If the engineering manager thinks that a questionable or potentially hazardous task order has been issued by upper management or a customer, that manager should make every effort to make known all aspects of that order that could potentially endanger safety, health, or the environment. In light of current events, suppose an upper manager asked an engineering team to develop software for diesel engines that would activate certain emission controls only during laboratory tests, not during actual use. Ideally, engineering managers should reject such projects even if it means losing challenging and advanced technology jobs for their subordinates.

Training should include classes on health and safety issues. Employees should be prepared to work under dangerous conditions. Some subordinates should be trained and certified in first aid and in emergency cardiovascular care. They should be recertified in fixed intervals. Employees being sent to offshore assignments should be trained in the customs, traditions, and languages of each country they will be working in, as people and their work habits change from country to country. Subordinates should have a firm grasp of internal and external workplace ethics relevant to each country. It is also very beneficial to learn some phrases in the native language (e.g., "good morning," "how are you?," "please," "thank you," "great job").

Training for offshore subordinates should take into account their background, including college degree, job experience, and other training, which can be much different than in the United States. University classes in foreign countries tend to emphasize theoretical knowledge over practical knowledge and provide only minimal hands-on practical experience. When foreign subordinates are not proficient in the English language, training must be given in their native language; even training notes and visual presentations must be translated into the native language.

Another important area of training for subordinates is the company's customers. Training should include discussions about each customer's organization and staff, as well as their needs and requirements for the company's products. The names of key contacts at the various customers should be provided and guidance given as to whom to approach at the customer when a technical question arises. A chain of communication for response to a customer's inquiry should be established, in addition to a chain of communication for when a subordinate needs to request something from a customer.

Subordinates must also be trained on the company's competitors—their organization and people, products, technological advantages and disadvantages, products under development. A chain of communication for response to a competitor's inquiry should be established, in addition to a chain of communication for when someone within the company wants to request something from a competitor.

Engineering managers must be aware of their team's technical competence levels and limitations. If the team is not qualified in a certain field, it might become necessary to bring in consultants or specialists in that particular field so the team can attack a technical task with competence and a high probability of success. Engineers should be encouraged to be active members of relevant professional engineering societies; for example, an engineer working in the field of robotics would benefit from being an active member of both the American Society of Mechanical Engineers (ASME) and the Institute of Electrical and Electronics Engineers (IEEE).

Engineering managers must be sure that everyone understands the company's priorities. When a design review deadline looms or if a customer is coming to the company for a crucial meeting, all other activities take on a lower priority, including advanced study classes.

Engineering managers should be very selective about sending subordinates to international conferences or technical shows. Such international trips can be very expensive and time consuming. It should be decided at least a year ahead of time (keeping in mind the department's travel budget) whether or not to send subordinates to foreign countries for further education or training. Upon their return, employees sent on such excursions should make a presentation to their department regarding the highlights of the international conference or technical show they attended.

PROTECTING THE COMPANY'S INTELLECTUAL PROPERTY

Engineering managers should always protect the company's intellectual property. Employees should be taught how to protect the company's intellectual property without antagonizing customers or competitors. Members of a team working on a classified project should be made aware of what exactly that means, and any potential conflicts of interest should be discussed. Other types of classes would include such topics as patent generation, patent documentation, patent lawyers, and processes for obtaining multinational patents.

Subordinates should always be encouraged to look for patentable ideas and should be reminded to write down their patentable findings in engineering books that are then signed and dated by colleagues. All contributors to a patentable idea should have their names included as inventors on a patent application. Patents increase the value of a company and give employees a sense of pride and extra spark to continue creating leading-edge technological inventions. Processing patent applications in foreign countries can take a long time and can be very expensive; the company and its patent lawyer must choose international patent application countries very carefully. Employees should have a good understanding of all patents that the company owns, as well as all patents that the company's competitors own.

Being published in a well-respected technical journal can be very gratifying, and subordinates should be encouraged to publish technical papers that do not violate the company's intellectual property. Employees should not spend too much time working on such endeavors, though. The author once had a subordinate who devoted almost 50% of his workday to publication-related issues, which was excessive. He was asked to spend at most 20% of his work time for publication-related issues and to do the rest of his publication work on his own time. It was difficult to make him focus more on his departmental tasks, but much persistence and micromanaging did the trick. He still published many technical papers, but he prioritized his departmental duties and put in an honest 32 hours of work for the department every week.

ANNUAL TRAINING PLANS

At the performance review meeting, the engineering manager and subordinate should agree on improvement areas for the next review period. After scheduling an employee's commitments to classes, conferences, shows, technical paper presentations, etc., the engineering manager can then assign departmental responsibilities accordingly. Typically, a subordinate's improvement activities should not exceed 400 hours per year, or 20% of 2000 annual work hours. Table 6.1 shows suggested improvements for an Engineer 2 during an annual review cycle. After attending conferences or trade shows, employees should always provide trip reports outlining their critical observations and findings. They should also present to the department highlights of the conference or trade show. Table 6.2 shows suggested improvements for a Technician 1 during an annual review cycle. Every course or training session should conclude with a final examination and certification. Subordinates should not be allowed to take courses or training classes that do not have a pass/fail requirement.

TABLE 6.1
Typical Suggested Improvements for an Engineer 2 During an Annual Review Cycle

List of Employee's Suggested Improvements for Year = 2016
Employee's Name: Tracy Salmon—Engr 2

Quarter	Course	Lecture Hours	Homework and Exam Hours
Q1	ASME Webinar—Shell and Tube Heat Exchanger Performance Testing	2	4
Q1	Consumer Electronics Show—Las Vegas, NV	24	48
Q1	Japanese Business Language and Business Customs—In house (Prof. Shinzo Ono)	26	52
Q2	Economics of Competing Energy—Stanford University online course	14	28
Q3	IEEE International Robotics Conference— Boston, MA	24	48
Q4	Industrial Statistics	24	48
	Total Hours	114	228

Total suggested improvement hours 342
Percentage of 2000 annual work hours 17.1%

TABLE 6.2
Typical Suggested Improvements for a Technician 1 During an Annual Review Cycle

List of Employee's Suggested Improvements for Year = 2016
Employee's Name: Nick Persie—Tech 1

Quarter	Course	Lecture Hours	Homework and Exam Hours
Q1	Surface Contamination Determination Technique Using Auger Electron Spectroscopy—AES Labs, Palo Alto, CA	24	48
Q2	MS Office—City College, San Jose, CA	26	52
Q3	Gauge Capability—In house	8	16
Q4	Industrial Statistics—In house	24	48
	Total Hours	82	164

Total suggested improvement hours 246
Percentage of 2000 annual work hours 12.3%

CHECKLIST FOR CHAPTER 6

AVOIDING OBSOLESCENCE

- An important part of the engineering manager's job is to ensure that the department's engineers and technicians continue their education and training to remain relevant in a changing world.
- Effective engineering managers will encourage their subordinates to maintain a separation between their work passions and their hobbies outside of work.
- Engineering managers should channel their subordinates' improvements in the right direction.

PLANNING FOR TRAINING

- Training should not normally be a secondary priority.
- Training for subordinates should extend beyond science and technology and cover world markets, environmental issues, legal issues, health issues, global customs and traditions, and, most importantly, your customers and competitors.
- Subordinates should receive the support necessary to complete simple tasks on time and with outstanding results.
- Ensuring the health and safety of subordinates should be a top priority.
- Under all conditions, engineering managers and their team members must maintain very high technical standards.
- Under no conditions should any type of bribery be accepted.
- Employees being sent to offshore assignments should be trained in the customs, traditions, and languages of each country they will be working in.
- Training for offshore subordinates should take into account their background, including college degree, job experience, and other training, which can be much different than in the United States.
- When foreign subordinates are not proficient in the English language, training must be given in their native language.
- Are subordinates well versed in the company's customer organizations and products?
- Are subordinates well versed in the company's competitor organizations and products?
- Engineering managers must be aware of their team's technical competence levels and limitations.
- Engineers should be encouraged to be active members of relevant professional engineering societies.
- Engineering managers must be sure that everyone understands the company's priorities.
- Engineering managers should be very selective about sending subordinates to international conferences or technical shows.

PROTECTING THE COMPANY'S INTELLECTUAL PROPERTY

- Employees should be taught how to protect the company's intellectual property without antagonizing customers or competitors.
- Members of a team working on a classified project should be made aware of what exactly that means.
- Classes should include such topics as patent generation, patent documentation, patent lawyers, and processes for obtaining multinational patents.
- Subordinates should always be encouraged to look for patentable ideas and should be reminded to write down their patentable findings in engineering books that are then signed and dated by colleagues.
- Employees should have a good understanding of all patents that the company owns, as well as all patents that the company's competitors own.
- Employees should be encouraged to publish technical papers that do not violate the company's intellectual property.

ANNUAL TRAINING PLANS

- At the performance review meeting, the engineering manager and subordinate should agree on improvement areas for the next review period.
- Typically, a subordinate's improvement activities should not exceed 400 hours per year, or 20% of 2000 annual work hours.
- After attending conferences or trade shows, employees should always provide trip reports outlining their critical observations and findings.
- The subordinate should also present to the department highlights of the conference or trade show.
- Every course or training session should conclude with a final examination and certification.
- Subordinates should not be allowed to take courses or training classes that do not have a pass/fail requirement.

7 Engineering Department Performance Reviews

PERFORMANCE REVIEW PREPARATION AND EXECUTION

Performance reviews are the most important function of an engineering manager. Performance reviews for every engineer and technician should take place at least once a year, and for new hires the first performance review should be held within the first 6 months of hiring date. Any time an issue with a subordinate becomes apparent, such as a project that requires improvement or a recognized deficiency in the subordinate's behavior, the engineering manager should discuss it immediately and not wait for performance review time. For serious problems, it is best to involve human resources in the discussion. Performance review cycles can differ from company to company and internationally; however, generally it is advisable to review the department's technical personnel at least once a year. Doing so will help steer team members in the right direction, for the benefit of the company and themselves.

Some engineering managers dread performance reviews and do not want to spend adequate time and effort preparing for them. They would prefer to get them over with as quickly as possible. Such an attitude, however, will hurt the department and company, in addition to the manager's subordinates, who will lose respect for their manager and the motivation to excel at their work.

Performance reviews should be one on one and face to face, whenever possible. A particularly conscientious engineering manager has been known to travel from the United States to South Korea to conduct an annual performance review with his resident engineer there rather than opting for teleconferencing or videoconferencing. This kind of above-and-beyond interaction between a boss and a subordinate fosters a strong bond, and the subordinate gains a sense of belonging to the group and to the company. Face-to-face interactions are not always possible for subordinates who work from home. It is possible for a manager to never actually meet such a subordinate. For such situations, teleconferencing or videoconferencing must be relied upon for the annual performance review.

Annual performance reviews should be held in a pleasant setting (e.g., small conference room), and an appropriate block of time should be set aside for the review. There should be no interruptions of any kind during performance reviews. If it is likely that a subordinate will be fired, a representative from human resources should be present at the meeting.

A typical performance review form is shown in Figure 7.1. Before filling out the performance review form, the engineering manager should seek input from other people that the subordinate has worked with professionally since the last review. These people can be from other departments in the company, customers, subcontractors, regulatory

Employee Name:	Date of Hire:
Current Job Title:	Date of Last Review:
Supervisor's Name and Title:	Present Review Date:
Employee's Present Job Duties:	

Job Knowledge: Knowledge of techniques, skills, processes and procedures of job
OUTSTANDING   HIGHLY EFFECTIVE   SATISFACTORY   NEEDS IMPROVEMENT   NEEDS MUCH IMPROVEMENT

Quality of Work: Ability to meet standards by accurate and careful work
OUTSTANDING   HIGHLY EFFECTIVE   SATISFACTORY   NEEDS IMPROVEMENT   NEEDS MUCH IMPROVEMENT

Quantity of Work: Ability to effectively use time and materials for high output of work
OUTSTANDING   HIGHLY EFFECTIVE   SATISFACTORY   NEEDS IMPROVEMENT   NEEDS MUCH IMPROVEMENT

Initiative: Takes action to improve work and learn new skills
OUTSTANDING   HIGHLY EFFECTIVE   SATISFACTORY   NEEDS IMPROVEMENT   NEEDS MUCH IMPROVEMENT

Climate Change: Takes action to improve environmental impact, sustainable development
OUTSTANDING   HIGHLY EFFECTIVE   SATISFACTORY   NEEDS IMPROVEMENT   NEEDS MUCH IMPROVEMENT

Adaptability: Ability to adjust to new situations, namely change
OUTSTANDING   HIGHLY EFFECTIVE   SATISFACTORY   NEEDS IMPROVEMENT   NEEDS MUCH IMPROVEMENT

Cooperation: Willingness and ability to work with others for mutual benefit
OUTSTANDING   HIGHLY EFFECTIVE   SATISFACTORY   NEEDS IMPROVEMENT   NEEDS MUCH IMPROVEMENT

Dependability: Reliability to complete jobs on time in a satisfactory manner
OUTSTANDING   HIGHLY EFFECTIVE   SATISFACTORY   NEEDS IMPROVEMENT   NEEDS MUCH IMPROVEMENT

Professionalism: Expertise in the manner and skill of performing one's job
OUTSTANDING   HIGHLY EFFECTIVE   SATISFACTORY   NEEDS IMPROVEMENT   NEEDS MUCH IMPROVEMENT

FIGURE 7.1   Sample performance review form.

agency representatives, or even a previous engineering department manager. When a subordinate has been reporting to a project manager, input should be sought from that project manager; also, if a subordinate has been on an offshore assignment, input should be sought from the appropriate managers within the offshore division.

Engineering managers should refrain from including any discriminatory remarks related to the subordinate's race, religion, age, sex, sexual orientation, disability, marital status, or any other type of personal or private matter. When a subordinate is demonstrating discipline problems or his work ethics are questionable, the engineering manager's written comments should be reviewed by a human resources expert.

Many external events can affect a department's performance. Morale can drop due to major declines in the stock market, customer order cancellations, delayed new product releases, upper management shake-ups, layoff rumors, travel freezes,

List Specific Achievements Since the Last Review:	
List Specific Problems Since the Last Review:	
List Employee's Strengths:	
List Employee's Suggested Improvement:	
Overall Performance Rating Comments:	
Overall Performance Rating: *Outstanding Highly Effective Satisfactory Needs Improvement Needs Much Improvement*	
Present Annual Salary:	Date of Last Annual Salary Raise:
Recommended Present Annual Salary Raise:	Amount of Last Annual Salary Raise:
Effective Date of Annual Salary Raise:	Last Promotion Date:
Recommended Bonus Amount:	Date of Last Bonus:
Recommended Stock Option Shares:	Date of Last Stock Option Rewarded:
Employee's Comments:	
Employee's Signature and Date:	Supervisor's Signature and Date:
Next Performance Review Date:	

FIGURE 7.1 (continued)   Sample performance review form.

budget cuts, no raises or bonuses being given, natural disasters, national disasters, international disasters, division shutdown, a death in the department or in the subordinate's family, a serious illness in the department or in the subordinate's family, etc. Whenever possible, annual performance meetings should be held in an environment that is calm and when morale is relatively good. If raises or bonuses are minimal or nonexistent, the engineering manager should be ready to explain the reasons why. In each review, the manager should be consistent in explaining why raises and bonuses are minimal or nonexistent.

The engineering manager should spend at least an average of 2 hours to prepare an engineer's performance review. The performance review begins with the engineer's current title and his or her current job duties, followed by a list all of the engineer's responsibilities (i.e., technical, administrative, professional) for the review period, with no ambiguities. This list of responsibilities should include those outside of the person's normal job duties (i.e., pet projects, sometimes referred to as side projects), as well as any assigned life-long learning projects designed to help the person avoid obsolescence and to broaden his or her technical knowledge base. Such personal projects should be of proven benefit to the employee and the company and should require no more than 10 to 20% of the person's weekly hours, depending on the company.

AREAS OF PERFORMANCE AND CONDUCT

As shown in Figure 7.1, the areas of performance and conduct include job knowledge, quality of work, quantity of work, initiative, consideration for environmental impact and sustainable development, adaptability, cooperation, dependability, and professionalism. These areas are evaluated as outstanding (9–10), highly effective (8–9), satisfactory (7–8), needs improvement (6–7), or needs much improvement (5–6). Although not shown in the figure, a number grade can be assigned, as indicated in the parentheses above. In each performance area, the engineering manager should include a couple sentences to explain why a particular rating was given. Some engineering managers do not give an outstanding (9–10) rating in the belief that no one can possibly achieve that highly sought-after status, but that is a wrong approach. Guidelines should be established for an engineer to achieve the highest possible evaluation rating level, and they should be encouraged to achieve that rating level in every aspect of their performance. Setting the bar high enhances the loyalty and morale of every engineer. Engineering managers should do what they can to help every subordinate achieve an outstanding rating level in every performance area. Inflating or deflating the performance ratings of subordinates is doing them no favors. Engineering managers who conduct their reviews in an honest and objective way will provide encouragement to their engineers to achieve an outstanding rating in every performance area.

Next on the list is the subordinate's strengths and suggested improvements. The engineering manager should be very clear about these items; they should be well defined and unambiguous. Future performance goals are goals that the subordinate should achieve during the next review period. These future performance goals should be reasonable so they do not discourage the subordinate but instead provide professional and personal motivation. The subordinate should always be involved in finalizing suggested improvement goals during the annual performance review meeting. A typical list of strengths and suggested improvements for a Level III structural design engineer is shown in Table 7.1.

All of the performance ratings are combined to derive an overall performance rating for the subordinate. Why a particular overall performance rating was given should be explained in writing. Again, the overall performance rating will be outstanding (9–10), highly effective (8–9), satisfactory (7–8), needs improvement (6–7), or needs much improvement (5–6).

You will never obtain the highest achievement levels for every subordinate, despite giving them a fair chance to correct any shortcomings. Such an engineering manager has not failed them. Detailed and conscientious performance reviews

TABLE 7.1

Level III Structural Design Engineer's Strengths and Suggested Improvements List for an Annual Performance Review

Employee's strengths for a Level III structural design engineer

1   Very proficient in linear and nonlinear stress analysis of our products using the in-house finite-element modeling software.

2   Excellent in modeling complicated geometries and boundaries of our products while minimizing weights, materials, and costs.

3   Modified the gear system design of our remotely operated underwater vehicle in three weeks. It is now operating without any failure.

4   Very timely and precise in calculation reports.

5   Very open and honest when encountering issues during finite-element modeling and seeks help from our consultants.

Employee's suggested improvements for a Level III structural design engineer

1   Take online course to improve ability in dynamic finite-element modeling (Advanced Analysis of Marine Structures, at Norwegian University of Science and Technology).

2   Attend a materials conference for stress analysis modeling at very low temperatures ($-50°C$) (International Conference on Composite Materials).

3   Mentor Alan Virtue, a Level I structural design engineer.

represent an opportunity for the engineering manager to mentally rank subordinates and to treat them accordingly, specifically with regard to assignments, salary raises, bonuses, and promotions. The subordinate's written comments should include how the subordinate has performed throughout the performance review period, any performance elements that could be improved upon, and future goals. During the annual performance review meeting, the engineering manager should always praise a subordinate's high performance ratings and provide positive input and encouragement for low or average performance ratings.

ANNUAL SALARY INCREASES AND BONUSES

Most companies provide a pool of funds for departmental salary increases and bonuses. The engineering manager must decide how to distribute this pool fairly among the subordinates. Each subordinate's overall performance rating and current salary must be considered to come up with just and fair increases. The calculations for typical salary increases and bonuses using both overall performance ratings and current salaries as guidelines are shown in Tables 7.2 and 7.3.

Before trying to distribute the annual salary increase and bonus pool to subordinates, the engineering manager must make sure that pool amounts allocated to the department are consistent with those for other departments (e.g., 3% salary increase pool for the engineering department vs. 10% for the sales department). Word about inconsistent salary and bonus pools will get out fast and can bring down a department's morale. The engineering manager must be adamant about receiving a fair share of the annual salary increase and bonus pools for the department.

TABLE 7.2

Typical Spreadsheet Calculations for Salary Raises Using 50% Overall Performance Ratings and 50% Current Salaries as Guidelines for an Engineering Department of Ten People

Annual salary increase pool percentage = 5%
Total annual salary increase pool = $47,524.75
Current salary weight = 50%
Overall performance rating weight = 50%

Employee	Current Annual Salary	Overall Performance Rating	Numerical Overall Performance Rating	Recommended Annual Salary Increase	Percentage Annual Salary Increase	New Annual Salary as of 1/1/2017
Technician 2	$52,000	Highly effective	8.5	$3825	7.36%	$55,825
Technician 3	$66,700	Highly effective	8.5	$4192	6.29%	$70,892
Engineer 1	$81,500	Highly effective	8.5	$4562	5.60%	$86,062
Engineer 1	$83,600	Satisfactory	7.5	$4318	5.16%	$87,918
Engineer 2	$103,000	Satisfactory	7.5	$4803	4.66%	$107,803
Engineer 2	$104,300	Needs improvement	6.5	$4538	4.35%	$108,838
Engineer 2	$104,950	Highly effective	8.5	$5149	4.91%	$110,099
Engineer 2	$106,115	Outstanding	9.5	$5475	5.16%	$111,590
Engineer 3	$121,880	Highly effective	8.5	$5572	4.57%	$127,452
Engineer 3	$126,450	Needs improvement	6.5	$5092	4.03%	$131,542
Total	$950,495	—	80.0	$47,526	—	—

TABLE 7.3

Typical Spreadsheet Calculations for Bonus Pool Distribution Using 75% Overall Performance Ratings and 25% Current Salaries as Guidelines for an Engineering Department of Ten People

Bonus pool = $100,000
Current salary weight = 25%
Overall performance rating weight = 75%

Employee	Current Annual Salary	Overall Performance Rating	Numerical Overall Performance Rating	Recommended Bonus	Percentage of Annual Salary
Technician 2	$52,000	Highly effective	8.5	$9336	17.95%
Technician 3	$66,700	Highly effective	8.5	$9723	14.58%
Engineer 1	$81,500	Highly effective	8.5	$10,112	12.41%
Engineer 1	$83,600	Satisfactory	7.5	$9230	11.04%
Engineer 2	$103,000	Satisfactory	7.5	$9740	9.46%
Engineer 2	$104,300	Needs improvement	6.5	$8837	8.47%
Engineer 2	$104,950	Highly effective	8.5	$10,729	10.22%
Engineer 2	$106,115	Outstanding	9.5	$11,697	11.02%
Engineer 3	$121,880	Highly effective	8.5	$11,174	9.17%
Engineer 3	$126,450	Needs improvement	6.5	$9422	7.45%
Total	$950,495	—	80.0	$100,000	—

Guidelines for annual salary increases and bonuses can vary from country to country. Engineering managers should ask their subordinates to be discreet about their annual salary increases and bonuses, particularly with regard to their international colleagues. The salary increase calculations shown in Table 7.2 assume a current salary weight of 50% and an overall performance rating weight of 50%, but these weights can vary from manager to manager. Table 7.3 presents an example of annual bonus distribution using a current salary weight of 25% and an overall performance rating weight of 75%. Recommended annual salary increases and bonuses are calculated as follows:

$$\text{Annual salary increase or bonus} = [W_{salary} \times (S/TS) + W_{performance} \times (R/TR)] \times P \quad (7.1)$$

where
  W_{salary} = Current salary weight.
  S = Current annual salary of a subordinate.
  TS = Total annual salary of the department.
  $W_{performance}$ = Overall performance rating weight.
  R = Numerical overall performance rating of a subordinate.
  TR = Total numerical overall performance rating.
  P = Total annual salary increase pool or bonus pool.

The sum of salary weight and performance weight should always be unity:

$$W_{salary} + W_{performance} = 1.0 \qquad (7.2)$$

Performance review ratings and comments, annual salary increases, and annual bonuses should not come as a complete surprise to subordinates, as the engineering manager should be discussing all performance areas during the weekly one-on-one meetings throughout the year. All of the interactions, discussions, and task assignments throughout the year should lead up to annual performance reviews without any surprises.

Sometimes during annual salary increase and bonus discussions, a subordinate might want to negotiate the proposed increase, threatening to resign if the desired annual salary raise and bonus are not given. In such cases, the engineering manager must stand firm. Such subordinates should be allowed to resign, unless they are badly needed for a crucial project, in which case the manager should negotiate but immediately begin looking for a replacement. There is no reason to keep subordinates who are not happy with the job, their manager, the department, or the company. They will only lower morale and perform less than optimally because they are focused on other opportunities outside of the company.

PROMOTIONS

The annual performance review meeting should include a discussion about the promotion status of the subordinate. Promotions should not be awarded simply for seniority but should reflect job knowledge, quality of work, quantity of work, initiative, consideration for environmental impact and sustainable development, adaptability, cooperation, dependability, and professionalism. The author has managed engineers who remained at Level 1 for the first several years of their professional careers. Promotions are not meant for everybody. Some subordinates are content with their current status and do not want to take on the additional responsibilities that come with advancement. The salaries of such subordinates can continue to increase into the upper levels of their salary range, which is appropriate for individuals who have devoted their professional lives to the company.

The engineering manager should not accelerate or decelerate promotions for deserving individuals. It can be very demoralizing for individuals to have a promotion delayed. Their performance level will drop, and they are likely to start looking for other opportunities. On the other hand, prematurely accelerating the promotion of individuals can cause them great stress when they encounter a more demanding environment.

Engineering managers should consider very carefully the initial salaries of new hires. Job knowledge and previous accomplishments should be taken into account when establishing the salary of any new subordinate. It is never advisable, for example, to hire an engineer with Level 2 job knowledge for a Level 3 position, because doing so will almost certainly bring down the morale of the other employees in the department.

The salary of a subordinate who is being promoted should be bumped up to the next level. Typical salary ranges for an engineering department are shown in Table 7.4. Engineering managers should check with the human resources department at least once a year to make sure that the department's salary ranges are compatible with others in the industry and location. Salary ranges can vary by engineering discipline, work location, and economic outlook. Salaries must be competitive in order to attract the highest caliber people. Engineering managers should not be surprised when a subordinate who has been trained and groomed for several years jumps ship with little notice.

TABLE 7.4
Typical Salary Ranges for Fiscal Year 2016

Employee Level	Salary Range
Technician 1	$40,000–$50,000
Technician 2	$50,000–$60,000
Technician 3	$60,000–$80,000
Senior Technician	$80,000–$100,000
Engineer 1	$80,000–$100,000
Engineer 2	$100,000–$120,000
Engineer 3	$120,000–$150,000
Senior Engineer	$150,000–$180,000

CHECKLIST FOR CHAPTER 7

PERFORMANCE REVIEW PREPARATION AND EXECUTION

- Performance reviews for every engineer and technician should take place at least once a year, and for new hires the first performance review should be held within the first 6 months of hiring date.
- Any time an issue with a subordinate becomes apparent, such as a project that requires improvement or a recognized deficiency in the subordinate's behavior, the engineering manager should discuss it immediately and not wait for performance review time.
- Performance reviews are preferably conducted one on one and face to face.
- For some situations, teleconferencing or videoconferencing must be relied upon for the annual performance review.
- Annual performance reviews should be held in a pleasant setting, and an appropriate block of time should be set aside for the review.
- There should be no interruptions of any kind during performance reviews.
- If it is likely that a subordinate will be fired, a representative from human resources should be present at the meeting.
- The performance review form should be fully filled out before the meeting with the subordinate.
- Before filling out the performance review form, the engineering manager should seek input from other people that the subordinate has worked with professionally since the last review.
- Engineering managers should refrain from including any discriminatory remarks related to the subordinate's race, religion, age, sex, sexual orientation, disability, marital status, or any other type of personal or private matter.
- When a subordinate is demonstrating discipline problems or his work ethics are questionable, the engineering manager's written comments should be reviewed by a human resources expert.
- Whenever possible, annual performance meetings should be held in an environment that is calm and when morale is relatively good.
- If raises or bonuses are minimal or nonexistent, the engineering manager should be ready to explain the reasons why.
- In each review, the manager should be consistent in explaining why raises and bonuses are minimal or nonexistent.
- The engineering manager should spend at least an average of 2 hours to prepare an engineer's performance review.
- The annual performance review form should be filled out completely.
- The performance review begins with the engineer's current title and current job duties, followed by a list all of the engineer's responsibilities (i.e., technical, administrative, professional) for the review period, with no ambiguities.
- This list of responsibilities should include those outside of the person's normal job duties (i.e., pet projects), as well as any assigned life-long learning projects designed to help the person avoid obsolescence and to broaden his or her technical knowledge base.

Areas of Performance and Conduct

- Subordinates should be rated for all areas of performance and conduct listed in the annual performance review form.
- The engineering manager should include a couple sentences to explain why a particular rating was given.
- Engineering managers should do what they can to help every subordinate achieve an outstanding rating level in every performance area.
- Inflating or deflating the performance ratings of subordinates is doing them no favors.
- The subordinate's strengths and suggested improvements should be listed. The engineering manager should be very clear about these items; they should be well defined and unambiguous.
- Future performance goals are goals that the subordinate should achieve during the next review period. These goals should be reasonable so they do not discourage the subordinate but instead provide professional and personal motivation.
- The subordinate should always be involved in finalizing suggested improvement goals during the annual performance review meeting.
- Why a particular overall performance rating was given should be explained in writing.
- The subordinate's written comments should include how the subordinate has performed throughout the performance review period, any performance elements that could be improved upon, and future goals.
- During the annual performance review meeting, the engineering manager should always praise a subordinate's high performance ratings and provide positive input and encouragement for low or average performance ratings.

Annual Salary Increases and Bonuses

- The engineering manager must decide how to distribute the pool of annual raises and bonuses fairly among the subordinates.
- Each subordinate's overall performance rating and current salary must be considered to come up with just and fair increases.
- When a subordinate wants to negotiate the proposed increase, threatening to resign if the desired annual salary raise and bonus are not given, the engineering manager must stand firm.
- Use of a spreadsheet such as Excel® can be helpful in determining a fair distribution of annual salary increases and bonus pools.

Promotions

- The annual performance review meeting should include a discussion about the promotion status of the subordinate.
- Promotions should not be awarded simply for seniority but should reflect job knowledge, quality of work, quantity of work, initiative, consideration

for environmental impact and sustainable development, adaptability, cooperation, dependability, and professionalism.

- The engineering manager should not accelerate or decelerate promotions
  for deserving individuals.
- Engineering managers should carefully consider the initial salaries of new
  hires.
- The salary of a subordinate who is being promoted should be bumped up
  to the next level.
- Engineering managers should check with the human resources department
  at least once a year to make sure that the department's salary ranges are
  compatible with others in the industry and location.

8 Laying Off, Firing, or Losing a Team Member

One of the most stressful tasks for an engineering manager is to lay off, fire, or otherwise lose a team member. No matter how many times an engineering manager has fired or laid off employees, it never becomes any easier. Engineering managers must work to maintain the morale of the team members left behind in the department. Layoffs and firings require very careful planning and quick execution, and it is advisable to have a human resources person present for each. It is also necessary to coordinate with the information technology department to cut off the former employee's digital access, and the security department should have a representative stationed nearby when the individual exits. Litigation involving the company and the engineering manager is always possible as a result of a layoff or firing. Engineering managers must be prepared for engineers and technicians leaving the department at any time, such as in the middle of a very important project or just before a critical technical presentation to a customer.

LAYING OFF A TEAM MEMBER

Layoffs are easier to manage than firing a team member. Because engineering and technician positions are generally "at will" employment, an employee can be terminated at any time and for any reason. Tough financial conditions, company downsizing, or shutting down a division can cause layoffs across a company. Engineering managers might get instructions from upper management or human resources to lay off, for example, 10% of the workforce or to cut 15% percent of the budget by a particular date. Some companies do layoffs at least once a year to eliminate low performers, troublemakers, or team members who are getting ready to leave the company anyway, but sometimes it is necessary to lay off good people, too.

When there is a layoff brewing on the horizon, the rumor mill runs at full speed. It is very difficult to keep everyone focused on their tasks until the actual layoff takes place. During this stormy and uncertain period, all team members must be given the same information about the layoff. An example response to the team can be "Yes, there is a layoff coming, but I do not know the extent of it or when it will happen."

Engineering managers should obtain details about layoffs and their timing from their supervisors and human resources department. These details must remain confidential until the layoffs occur. Engineering managers should be familiar with any career counseling, retraining strategies, outplacement sources, etc., that might be available to the people being laid off. They must also understand the compensation packages that will be handed out.

When deciding which team members to let go, the engineering manager should first consider low performers, troublemakers, team members who are getting ready to leave the company anyway, and team members who do not fit the department's future plans. Toughest to lay off are team members who have contributed to the department for many years but do not fit into the department's future plans anymore. The job performance of some seasoned team members can decline through the years, as they become content with the status quo and do not care about advancement. These seasoned team members can be close friends of the manager and perhaps mentored several novice engineers and technicians. After developing a list of employees to be laid off, the engineering manager should share that list with their supervisors for approval. It must then be presented to human resources, who will strategize in detail the timing of the layoffs. As noted earlier, layoffs must also be coordinated with the information technology and security departments.

The information technology department will disable the company e-mail address and the desktop computer of the person being laid off when directed to do so by the engineering manager. This will help prevent the person from sending out a scathing companywide e-mail. The engineering manager will give the security person a list of items that must be collected from the person being laid off before he or she leaves the premises. Such a list might include keys, laptops, cellphones, or document-controlled engineering notebooks. The security person will escort the person being laid off from the engineering manager's office to the employee's office so he or she can collect personal items, turn over any items on the list provided to security, and say goodbye to friends and colleagues. The person will then be escorted out the door by security, who will disable the person's entry codes to company buildings or laboratories.

A layoff meeting should not last more than 10 to 15 minutes. Most layoffs are done on Thursday or Friday mornings. If the engineering manager has, say, five team members to lay off, the entire process should be completed within an hour or at most an hour and a half. It is strongly recommended that a human resources person be in the engineering manager's office to provide support during these layoff meetings. Generally, the human resources person will discuss final paychecks, 401(k) benefits, stock options, and health insurance, among other things. Engineering managers should be well prepared for layoff meetings. They should control their anxiety and not deviate from the planned layoff routine. Those being laid off should be called one by one to the manager's office. A typical layoff conversation might go as follows: "Hi, Alan. Please sit down. I am sorry to inform you that I am laying you off today due to our company's downsizing. You have contributed to this department immensely. Good luck in your future endeavors. Karen from our human resources department will go over the details with you. A security officer will escort you out of the building after you have gathered your personal items and hand over to him any company items that you possess."

It is best to allow team members being laid off to express their feelings. Alan might become tearful or angry, or he might accept his fate, listen to the human resources representative, and leave your office without saying a word. A person who becomes very angry might have to be escorted immediately out of the building by security officers. Ideally, a laid-off person might express his or her feelings as follows: "I am sorry to learn that I am being laid off. I had a good career under your supervision and leadership. I am sure I will find another job soon. Can I get a letter of recommendation from you?" The manager should answer, "Of course, Alan. Please let me know your personal e-mail and I will send you one by the end of next week."

Unforeseen complications could arise during the layoff process. A person to be laid off might happen to call in sick that day, might be on a business trip, or might be on vacation or taking a personal day. How to handle such cases should be discussed with the human resources department.

Gossip about the layoff will slow down after it actually occurs. Team members must try to return to their tasks, and the engineering manager should do everything possible to re-motivate the team. A team meeting should be called right away so the engineering manager can explain truthfully to the team why the layoff was necessary and clear up any misunderstandings. The engineering manager must be able to light a fire under the team again. Sometimes it might help to bring in the company's president or other high-level executive to help answer questions that might arise regarding the future of the company and job security.

The author was affected by layoffs when the dot-com bubble burst in 2001. Employees got the bad news that our corporation was going to close down the division's operations. We were a data communication chip design group of about 70 people composed of engineers and project managers. The team's efficiency immediately suffered. Everyone was worrying about their future rather than focusing on their tasks. This unfortunate situation lasted for 10 days, at which point a group of executives from corporate came to the facility and finalized layoff details with the division's general manager. That same day, all of us were summoned to a large conference room, where we learned about how the division shutdown process was going to take place. It would be a staggered layoff process depending on the status of the various projects. In the first week 50% of the total workforce was laid off. Everyone received compensation packages consisting of vacation pay and bonuses (typically one month's salary for every year of service), depending on seniority and salary level. My team of four engineers was given 3 months to finish our project and get it approved by our customer. My team and I worked very closely during those final 3 months. I allowed my team members to go to job interviews during work hours as long as they made up the lost time later. Our offices were deserted. During the phase-out process, we were saying goodbye to several colleagues every week. My team and I remained close and discussed our future opportunities and plans. We were able to finish our project successfully even knowing that the end was near. This episode of division shutdown was less painful than it might have been. We knew what we had to do and when the end was, and we were all able to secure new jobs within those 3 months.

FIRING A TEAM MEMBER

It is much more difficult to fire an employee than to lay off an employee. Reasons for firing a team member cannot give any indication of race, gender, religion, or age involvement. A team member's termination process must be documented meticulously in case a lawsuit results from the termination. A team member's performance deficiency or unacceptable behavior should be presented to him or to her in a written document. This performance review document should be prepared carefully and should focus only on the employee's work performance deficiency or behavior issues. The performance review document should include a work performance or behavior improvement plan and timeline for the plan. Engineering managers should discuss the problem employee and the performance review document with their supervisor for approval to proceed. A representative from human resources should also review the document.

When the performance review document is ready to be presented to the team member, a human resources representative should be present at the meeting, which should be short and to the point. During the meeting, any agreed-upon changes to the work performance or behavior improvement plan and timeline should be noted, and the performance document should be updated and signed by all parties. This document should then go into the employee's personnel file. During this meeting, the engineering manager should encourage the team member to follow the improvement plan. The team member must realize that a lack of acceptable improvement in performance or behavior could result in termination.

If the warned team member does not improve his or her performance or behavior within the time allowed by the improvement plan, the engineering manager can prepare a second and final work performance document, which essentially offers the employee a second chance. The meeting to present this document should proceed the same as that for the first work performance document. The team member should come out of this performance review meeting understanding that this is his or her last chance. Many engineers and technicians will work to improve their performance or behavior in order not to be terminated, although some performance degradation and unacceptable behavior might be caused by issues outside of work, such as family problems, financial issues, or substance abuse, which can be difficult to overcome.

If things do not improve after the second warning, the engineering manager should not procrastinate and should terminate the team member immediately. After preparing a termination document, the engineering manager should discuss it with the employee's supervisor and with the human resources department. The termination process should involve the human resources department, information technology department, and security department, similar to the layoff process detailed earlier. During the termination meeting, the engineering manager should focus only on the reasons for the termination. During the termination meeting, the terminated employee should be paid all salary owed up to and including the termination date, in addition to any accrued vacation time. It is not necessary to provide severance pay. The termination document should be signed by all parties and added to the terminated team member's personnel file. The termination meeting should be short and to the point, and a human resources representative should be present.

The information technology department will disable the company e-mail address and the desktop computer of the person being terminated when directed to do so by the engineering manager. The security person will escort the person being laid off from the engineering manager's office to the employee's office so he or she can collect personal items, turn over any items on the list provided to security, and say goodbye to friends and colleagues. The person will then be escorted out the door by security, who will disable the person's entry codes to company buildings or laboratories. Soon after the termination process is over, the engineering manager should hold a team meeting to explain to team members why their colleague was terminated. The engineering manager should emphasize the fact that the terminated colleague was given several opportunities to improve.

Suppose subordinates begin to complain about a colleague: "I am working my butt off, but Sam is on the phone all day with his girlfriend. And when he's not on the phone he's playing computer games and doing very little work." An engineering manager who gets complaints about a team member from several subordinates should take immediate action before the team's performance declines and good people begin to leave.

LOSING A TEAM MEMBER

An engineering manager can spend a lot of time, effort, and funds to groom an engineer or a technician who one day comes into work and announces, "Hi, boss, I'm quitting today." Sometimes the engineering manager will be taken totally by surprise, sometimes not. Engineering managers must be ready for personnel losses, as the days of lifetime employment are gone. An engineer stays in a position for an average of only about 3 years. The author has held eight different engineering jobs during his 36 years of working. The shortest job lasted 2 months and the longest was 17 years. Reasons for switching jobs included obtaining a better position or higher pay, as well as a shorter commute.

When a subordinate suddenly quits, the engineering manager can seek temporary help from among the department's network of consultants or from the global pool of engineering freelancers. Stealing an engineer from a competitor or hiring and grooming a novice engineer can take months. An unfortunate situation occurs when the workload of the subordinate who quit must be distributed to the remaining subordinates. Their stress levels can climb quickly.

Engineers and technicians can quit for several different reasons. The most common reasons are that they do not like their manager or the company's leadership, they do not like where the company is heading, they are not challenged enough by their assignments, or they are not satisfied with their pay or other benefits. Some might realize that engineering is not meant for them, and they want to experience another professional field, such as marketing, finance, law, or medicine. Engineers can quit to become entrepreneurs so they can further develop concepts ranging from vehicle crashworthiness to renewable energies. Rather than leaving the company altogether, some subordinates might be good candidates to grow within the company and move to other departments as leads or as managers, either locally or internationally.

Whatever the case, engineering managers must be ready to deal with engineers and technicians who quit with might seem only a moment's notice. Some subordinates will not be courteous enough to give the engineering manager two-weeks' notice. Engineering managers must have a good personnel plan that involves extensive cross-training. They should also have a solid network of consultants and a reliable pool of global engineering freelancers.

CHECKLIST FOR CHAPTER 8

LAYING OFF A TEAM MEMBER

- All team members must be given the same information about a layoff.
- Engineering managers should obtain details about layoffs and their timing from their supervisors and human resources department.
- Details about a layoff must remain confidential until the layoff occurs.
- Engineering managers should be familiar with any career counseling, retraining strategies, outplacement sources, etc., that might be available to the people being laid off.
- Engineering managers must also understand the compensation packages that will be handed out.
- When deciding which team members to let go, the engineering manager should first consider low performers, troublemakers, team members who are getting ready to leave the company anyway, and team members who do not fit the department's future plans.
- After developing a list of employees to be laid off, the engineering manager should share that list with their supervisors for approval.
- This list must then be presented to human resources, who will strategize in detail the timing of the layoffs.
- You have to also coordinate the finalized layoff process with your information technology and with your security departments.
- A human resources person should be present during layoff meetings.
- The information technology department will disable the company e-mail address and the desktop computer of the person being laid off when directed to do so by the engineering manager.
- The engineering manager will give the security person a list of items that must be collected from the person being laid off before he or she leaves the premises.
- The security person will escort the person being laid off from the engineering manager's office to the employee's office so he or she can collect personal items, turn over any items on the list provided to security, and say goodbye to friends and colleagues.
- The person will then be escorted out the door by security, who will disable the person's entry codes to company buildings or laboratories.
- A layoff meeting should not last more than 10 to 15 minutes.
- Most layoffs are done on Thursday or Friday mornings.
- Engineering managers should be well prepared for layoff meetings.
- Engineering managers should control their anxiety and not deviate from the planned layoff routine.
- Those being laid off should be called one by one to the engineering manager's office.
- Unforeseen complications could arise during the layoff process, such as a person to be laid off calls in sick that day, is on a business trip, or is on vacation or taking a personal day. How to handle such cases should be discussed with the human resources department.

- A team meeting should be called right away so the engineering manager can explain truthfully to the team why the layoff was necessary and clear up any misunderstandings.
- The engineering manager must be able to light a fire under the team again.
- Sometimes it might help to bring in the company's president or other high-level executive to help answer questions that might arise regarding the future of the company and job security.

FIRING A TEAM MEMBER

- It is much more difficult to fire an employee than to lay off an employee.
- Reasons for firing a team member cannot give any indication of race, gender, religion, or age involvement.
- A team member's termination process must be documented meticulously in case a lawsuit results from the termination.
- A team member's performance deficiency or unacceptable behavior should be presented to him or to her in a written document.
- The performance review document should include a work performance or behavior improvement plan and timeline for the plan.
- Engineering managers should discuss the problem employee and the performance review document with their supervisor for approval to proceed.
- A representative from human resources should also review the document.
- When the performance review document is ready to be presented to the team member, a human resources representative should be present at the meeting, which should be short and to the point.
- During the meeting, any agreed-upon changes to the work performance or behavior improvement plan and timeline should be noted, and the performance document should be updated and signed by all parties.
- This document should go into the employee's personnel file.
- During this meeting, the engineering manager should encourage the team member to follow the improvement plan. The team member must realize that a lack of acceptable improvement in performance or behavior could result in termination.
- If the warned team member does not improve his or her performance or behavior within the time allowed by the improvement plan, the engineering manager can prepare a second and final work performance document, which essentially offers the employee a second chance.
- The meeting to present this document should proceed the same as that for the first work performance document. The team member should come out of this performance review meeting understanding that this is his or her last chance.
- If things do not improve after the second warning, the engineering manager should not procrastinate and should terminate the team member immediately.
- After preparing a termination document, the engineering manager should discuss it with the employee's supervisor and with the human resources department.

- The termination process should involve the human resources department, information technology department, and security department, similar to the layoff process detailed earlier.
- During the termination meeting, the engineering manager should focus only on the reasons for the termination.
- During the termination meeting, the terminated employee should be paid all salary owed up to and including the termination date, in addition to any accrued vacation time. It is not necessary to provide severance pay.
- The termination document should be signed by all parties and added to the terminated team member's personnel file.
- The termination meeting should be short and to the point, and a human resources representative should be present.
- The information technology department will disable the company e-mail address and the desktop computer of the person being terminated when directed to do so by the engineering manager.
- The security person will escort the person being fired from the engineering manager's office to his or her office, where the former employee can collect personal items, turn over any items on the list provided to security, and say goodbye to friends and colleagues.
- The person will then be escorted out the door by security, who will disable the person's entry codes to company buildings or laboratories.
- Soon after the termination process is over, the engineering manager should hold a team meeting to explain to team members why their colleague was terminated. The engineering manager should emphasize the fact that the terminated colleague was given several opportunities to improve.

LOSING A TEAM MEMBER

- Engineering managers must be ready for personnel losses, as the days of lifetime employment are gone.
- An engineer stays in a position for an average of only about 3 years.
- An unfortunate situation occurs when the workload of the subordinate who quit must be distributed to the remaining subordinates. Their stress levels can climb quickly.
- Engineering managers must have a good personnel plan that involves extensive cross-training. They should also have a solid network of consultants and a reliable pool of global engineering freelancers.

9 Essential Engineering Tools and Workplace Environment

ESSENTIAL ENGINEERING TOOLS

Engineering tools have evolved continuously throughout human existence. The author has witnessed, over the last 60 years, an exponential increase in engineering computing power and speed. Engineering computing tools have evolved from pencil and paper and a good eraser to slide rules, to punched cards and mainframe computers, to scientific calculators, to personal computers, to powerful desktop workstations, and finally to cloud computing featuring rapid access, flexibility, security, and low-cost servers with a broad set of interactive application services over the Internet. Also, computer programming languages have evolved from Fortran to JavaScript. The exponential increase in computer speed and storage capacity, coupled with extensive engineering knowledge about the physical and chemical properties of materials, has allowed engineers to derive optimal solutions to complicated, interactive, iterative, and nonlinear engineering problems.

This revolution in engineering calculations has brought with it some issues that must be controlled very strictly by an engineering manager. One of these issues is software version control. Versions of software such as AutoCAD® for computer-aided design; COMSOL Multiphysics® for modeling electrical, mechanical, fluid, and chemical applications; Microsoft® Excel®; and Microsoft® Project, as well as in-house developed software, must be strictly controlled. The software version control system should extend beyond the engineering department to the company's other departments, consultants, international facilities, customers, and subcontractors. Incompatible software versions can cause unnecessary errors and misinterpretations of engineering calculations and presentations. In some cases, the customer or subcontractor might not be able to open and review an engineering file.

The author has experienced not only software version incompatibility but also software incompatibility when a customer used an internally developed project management software rather than the popular MS Project software used for the firm's project management. Weekly project update reports to this particular customer had to be converted to PDF files so the customer's project manager could open and review them. This situation should have been resolved early on by specifying the project management software and its version to be used by all parties involved in the project.

The engineering manager must have very high standards for the development of user manuals and version controls for software developed in-house and for customer-developed software. It is not unusual for engineering software developed in-house to not have any user manuals or version controls. Engineers who are being trained in such software or have questions about it must chase down the software originator for help. Even the engineer who developed the software may not be able to explain the source of a constant that he added to fit his equation to experimental data. If the in-house software originator leaves the company without properly documenting the software, the engineering manager must assign someone else to spend countless hours deciphering the software to make it usable.

The revolution in engineering calculations has brought with it still other issues that must be handled by the engineering manager. One such issue is software training and keeping every engineer and technician up to date with the latest software versions being used in the department and at international facilities. Another issue is to have enough software protection keys (dongles) in the department for parallel computing; for example, the department's engineers might need several workstations and dongles to perform parallel finite-element modeling runs for a structural design optimization. Another issue is an engineering group's software training being compatible with a customer's requirements. The author recalls when his engineers were trained to design and verify a digital circuit in the Verilog hardware description language, but an international customer wanted their digital circuit to be designed and verified in VHDL. The circuit design engineers had to be trained in VHDL very quickly.

Security and computer file accessibility issues must be strictly controlled by engineering managers. An engineer could quit the department under adverse conditions and not provide crucial software files before leaving. The engineering manager and information technology department should be able to access the disgruntled employee's computer files through daily backups of his workstation and personal computer, without the use of passwords. Employees who leave the department should not be able to take confidential and technical company information with them. The engineering manager must take all necessary steps to protect the company's intellectual property with regard to software applications and data files.

The revolution in engineering calculations has been coupled with a revolution in engineering measurement systems. As the technology evolved so did the measurement standards and the measurement equipment; for example, the standard unit of length, the meter, evolved from the length of a prototype platinum–iridium bar to the wavelength of light that travels in a vacuum over a time interval of $1/299,792,458$ of a second. The standard unit of time, a second, evolved from a fraction of a mean solar day ($1/(24 \times 3600)$), which had irregularities due to Earth's rotation, to a transition between two energy levels of a cesium-133 atom at a frequency of $9,192,631,770$ hertz. Length measurement equipment evolved from measuring tapes and calipers to high-precision digital laser systems. Electrical measurement systems such as multimeters and spectrum analyzers have become more precise and accurate. Engineering managers must make sure that their subordinates have access to all of the latest measurement gadgets in their laboratories and toolkits. This collection of the latest measurement gadgets must be calibrated periodically and their calibrations traceable to international standards.

Engineers should have access to technical libraries in their fields of expertise. In addition to online sources of technical knowledge, engineers should have memberships in pertinent technical societies such as the American Society of Mechanical Engineers (ASME) and the Institute of Electrical and Electronics Engineers (IEEE). An engineering department's own technical library should be filled with technical journals, standards (national and international), handbooks, and technical books in the department's field of expertise. Every mechanical engineer should have a copy of the latest edition of *Marks' Standard Handbook for Mechanical Engineers*. Every chemical engineer should have a copy of the latest edition of *Perry's Chemical Engineers' Handbook*. An engineering department that is designing and building structures for offshore oil and gas field development should have the latest editions of the American Bureau of Shipping (ABS) and Det Norske Veritas (DNV) standards in the department. Copies of all relevant patents, especially competitors' patents, should be available in the technical library.

PRODUCTIVE ENGINEERING DEPARTMENT ENVIRONMENT

Engineers and technicians can have all of the latest gadgets and software, and they can have all of the latest technical information at their fingertips; however, the work environment must be conducive to keeping them stimulated, creative, and self-driven. Promoting a team environment is important. Every member of an engineering department must support each other, and brainstorming among subordinates should be encouraged. The engineering manager must do everything possible to maintain a relaxing, comfortable, and quiet environment.

Some engineering departments have made dramatic changes to keep their engineers happy and productive. Fully equipped exercise gyms and locker rooms, group tai chi classes, in-house masseuses, resting cots, and free lunches are becoming the norm. Dress codes, too, are evolving; for example, male engineers are abandoning suits, ties, and dress shoes in favor of casual blue jeans, t-shirts, and flip-flops. International-based engineering departments tend to have a more rigid and structured workplace compared to departments in the United States; for example, one is not likely to see a male engineer at work without a suit and a tie in Japan or South Korea. Also, in Japan and in South Korea subordinates generally do not head home before the department manager has left.

The ideal engineering environment promotes brainstorming, provides a sense of empowerment, and offers flexible hours. Engineers and technicians who abuse such a workplace environment require close supervision. Introverted team members may not like participating in brainstorming and empowerment discussions. Some subordinates would prefer to work only from 8 to 5, and some might prefer not to travel internationally. In an ideal engineering environment, the engineering manager would recognize each subordinate's differences and treat each of them accordingly.

Being able to keep stress at a tolerable level contributes greatly toward creating an ideal engineering environment. A good engineering manager will be flexible with regard to task completion time constraints and make every effort to minimize firefighting through the use of good project planning and sufficient manpower. A good engineering manager generates creativity and promotes a can-do attitude for a variety of tasks, ranging from menial to complex. If some of an engineering manager's team members are staying late or working weekends to prepare for a crucial design review or for preparation of a technical proposal, a good engineering manager will do the same, offering as much support and encouragement as possible (e.g., bringing in lunches or dinners, promising days off or a good bonus, offering small tokens such as ballgame tickets or golf course passes). Every assignment that is completed on time and correctly should be praised. The ultimate responsibility of engineering managers should be to prevent burnout by not overloading their subordinates.

If errors are made in the course of an assignment, employees should not be punished. Every engineer or technician should have the opportunity to make a mistake or two. All errors should be discussed in weekly engineering meetings so they serve as lessons learned for every subordinate. The engineering manager should make sure that all engineering drawings, calculations, and documents are checked independently and are properly document controlled.

Engineering managers should challenge their subordinates by continuing to expand their knowledge base by giving them varied and increasingly difficult assignments. These assignments can even include tasks outside of engineering, such as in manufacturing, sales and marketing, quality control, customer service, and dealing with subcontractors and regulatory agencies. Subordinates' horizons can be expanded by giving them international assignments, but they must be trained and properly prepared for such complicated tasks.

In a stimulating engineering environment, technical courses and conferences play an important role. They should be scheduled for each subordinate well in advance, and attendance should be recorded in the subordinates' formal reviews. The engineering manager should encourage subordinates to submit technical papers to technical journals and conferences without violating the company's intellectual properties.

Subordinates should also be encouraged to record any new and patentable ideas in their engineering notebooks. The company's patent attorneys can help them apply for patents. When a patent is granted, a congratulatory plaque should be presented at a recognition dinner. Subordinates should be given some freedom to work on patent applications and write technical papers. Some companies allow about 10% to 15% of an engineer's work time to be spent on his or her own engineering ideas.

A crucial factor in creating a stimulating engineering environment is ease of communication. All subordinates should have fast, easy access to e-mails, texting, teleconferencing, and videoconferencing. The network of communication should be available to all department personnel worldwide, with appropriate security controls in place. All communications should be recorded and important ones document controlled. All subordinates should be aware of international time zones and when their counterparts in, say, India or Germany can best be reached.

Another critical requirement for a stimulating engineering environment is conducting effective and well-controlled meetings. Many hours are spent in engineering meetings, and much of that time is wasted. Meeting organizers must make sure that only people involved with the subject matter are called into meetings. If a meeting will be long and cover several topics, the agenda should be divided into various segments such that those involved with a particular topic can participate in just that part of the meeting. A subordinate who misses a meeting should be able to access meeting minutes through document control to review them. Efficient meetings do not waste precious work hours. Subordinates should be fully informed and prepared for their part in a meeting.

Engineers and technicians can become very perturbed when a project specification suddenly changes or when a project is shelved without warning after they have spent months or years working on it. This is a way of life in engineering. A customer calls right before the vehicle prototyping phase and wants to increase the ground clearance of the vehicle. Or, a customer provides the wrong interface dimensions for a piece of equipment, which must then be redesigned when the correct dimensions are provided. When a project is shelved, engineering managers should emphasize the positive by highlighting accomplishments achieved during work on the project and pointing out the experience gained from working on it. Subordinates should understand that a change in specifications results in cost increases and an extended timeline for completing the project. Specification changes cannot be written on a napkin or transmitted by word of mouth. All specification changes must go through the proper channels of engineering change orders, signatures, and document control.

Engineering departments cannot live in a cocoon. Everyone in the department must interact with other departments within the company, both domestically and internationally. Subordinates will have to deal with customers, subcontractors, consultants, regulatory agencies, etc., so they must be groomed with regard to how to handle these various types of interactions. Subordinates must learn that it is not acceptable to make task commitments on someone else's behalf. They must become familiar with every department's pecking order so they know how to seek the appropriate help to complete a task. Bottom line is that everyone must learn how to be team members and how to operate within a multidisciplinary environment. Engineering managers should take the lead in training their subordinates how to deal with international partners, customers, subcontractors, and even stockholders. Everyone must understand that, to prevent insider trading, there is certain information that they cannot divulge (e.g., the release date of a new product) to stockholders or the public.

ISO 9001 AND ISO 14001

Almost every engineering company or organization is ISO 9001 and ISO 14001 certified, and every engineering company or organization has its own quality management system and continuous process improvement goals that cover all segments of a company or organization. Examples of an engineering department's processes that are covered by ISO 9001 and by ISO 14001 include document control procedures; change order procedures; training procedures; laboratory equipment calibration procedures; laboratory equipment precision and accuracy determination procedures; error control and elimination procedures; lessons-learned procedures; customer communication procedures; in-house (domestic and international) communication procedures; regulatory requirement procedures; product design, development, review, verification, and validation procedures; and nonconforming product control procedures, among others. All process procedures and their records must be documented and accessible to all subordinates. It is the engineering manager's responsibility to make sure that all subordinates are familiar with ISO 9001 and ISO 14001 quality and environmental management procedures and process improvement goals.

The engineering manager must adhere closely to the company's quality control program—namely, ISO 9001—and work to make continual process improvements. Subordinates should be encouraged to provide input regarding process improvement that does not deviate from the company's quality control procedures. The same rules apply to the company's environmental management system and ISO 14001. Subordinates should become involved in climate change mitigation, reducing energy and water consumption, waste management, etc. They should learn the company's environmental management system guidelines and understand their effects on the environment. For example, each subordinate should realize that using a ton of recycled paper rather than new saves about 17 trees. The engineering department's quality and environmental management systems should be the structural backbone of the subordinates' activities. Continual improvements will provide the best possible engineering workplace environment in the field.

CHECKLIST FOR CHAPTER 9

ESSENTIAL ENGINEERING TOOLS

- Engineering managers must be sure that the latest computer technology is available to their subordinates.
- Subordinates should be trained in the latest versions of computer programming languages.
- Strict version control must be maintained for all software in an engineering department.
- Strict version control must be maintained for all software that is shared with other departments in the company.
- Strict version control must be maintained for all software used by international facilities.
- Strict version control must be maintained for all software shared among customers, subcontractors, and consultants.
- Descriptive and up-to-date user manuals must be maintained for all software developed in-house and by customers.
- Every engineer and technician must be kept up to date with the latest software versions being used in the department and at international facilities.
- The engineering department should have sufficient software protection keys (dongles) for parallel computing.
- The engineering group's software training must be compatible with a customer's requirements.
- The engineering manager and information technology department should be able to access an employee's computer files through daily backups of his workstation and personal computer, without the use of passwords.
- The engineering manager must take all necessary steps to protect the company's intellectual property with regard to software applications and data files.

- Engineering managers must make sure that their subordinates have access to all of the latest measurement gadgets in their laboratories and toolkits.
- The department's collection of measurement gadgets must be calibrated periodically and their calibrations traceable to international standards.
- Engineers should have access to technical libraries in their fields of expertise.
- Engineers should have memberships in pertinent technical societies such as the American Society of Mechanical Engineers (ASME) and the Institute of Electrical and Electronics Engineers (IEEE).
- An engineering department's own technical library should be filled with technical journals, standards (national and international), handbooks, and technical books in the department's field of expertise.
- Every mechanical engineer should have a copy of the latest edition of *Marks' Standard Handbook for Mechanical Engineers.*
- Every chemical engineer should have a copy of the latest edition of *Perry's Chemical Engineers' Handbook.*
- Copies of all relevant patents, especially competitors' patents, should be available in the technical library.

PRODUCTIVE ENGINEERING DEPARTMENT ENVIRONMENT

- Promoting a team environment is important.
- Every member of an engineering department must support each other, and brainstorming among subordinates should be encouraged.
- The engineering manager must do everything possible to maintain a relaxing, comfortable, and quiet environment.
- The engineering department should have a stimulating work environment.
- International-based engineering departments tend to have a more rigid and structured workplace compared to departments in the United States.
- The ideal engineering environment promotes brainstorming, provides a sense of empowerment, and offers flexible hours. Engineers and technicians who abuse such a workplace environment require close supervision.
- In an ideal engineering environment, the engineering manager recognizes each subordinate's differences and treats each of them accordingly.
- Being able to keep stress at a tolerable level contributes greatly toward creating an ideal engineering environment.
- A good engineering manager will be flexible with regard to task completion time constraints and make every effort to minimize firefighting through the use of good project planning and sufficient manpower.
- A good engineering manager generates creativity and promotes a can-do attitude for a variety of tasks, ranging from menial to complex.
- If some of an engineering manager's team members are staying late or working weekends to prepare for a crucial design review or for preparation of a technical proposal, a good engineering manager will do the same, offering as much support and encouragement as possible.
- Every assignment that is completed on time and correctly should be praised.

- The ultimate responsibility of engineering managers should be to prevent burnout by not overloading their subordinates.
- If errors are made in the course of an assignment, employees should not be punished.
- All errors should be discussed in weekly engineering meetings so they serve as lessons learned for every subordinate.
- Engineering managers should challenge their subordinates by continuing to expand their knowledge base by giving them varied and increasingly difficult assignments.
- Subordinates' horizons can be expanded by giving them international assignments, but they must be trained and properly prepared for such complicated tasks.
- Technical courses and conferences should be scheduled for each subordinate well in advance, and attendance should be recorded in the subordinates' formal reviews.
- The engineering manager should encourage subordinates to submit technical papers to technical journals and conferences without violating the company's intellectual properties.
- Subordinates should also be encouraged to record any new and patentable ideas in their engineering notebooks.
- The company's patent attorneys can help subordinates apply for patents.
- When a patent is granted, a congratulatory plaque should be presented at a recognition dinner.
- Subordinates should be given some freedom to work on patent applications and write technical papers.
- All subordinates should have fast, easy access to e-mails, texting, teleconferencing, and videoconferencing.
- The network of communication should be available to all department personnel worldwide, with appropriate security controls in place.
- All communications should be recorded and important ones document controlled.
- All subordinates should be aware of international time zones and when their counterparts in a foreign country can best be reached.
- Meeting organizers must make sure that only people involved with the subject matter are called into meetings.
- If a meeting will be long and cover several topics, the agenda should be divided into various segments such that those involved with a particular topic can participate in just that part of the meeting.
- A subordinate who misses a meeting should be able to access meeting minutes through document control to review them.
- Subordinates should be fully informed and prepared for their part in a meeting.
- When a project is shelved, engineering managers should emphasize the positive by highlighting accomplishments achieved during work on the project and pointing out the experience gained from working on it.

- Specification changes cannot be written on a napkin or transmitted by word of mouth. All specification changes must go through the proper channels of engineering change orders, signatures, and document control.
- Subordinates will have to deal with customers, subcontractors, consultants, regulatory agencies, etc., so they must be groomed with regard to how to handle these various types of interactions.
- Subordinates must learn that it is not acceptable to make task commitments on someone else's behalf.
- Subordinates must become familiar with every department's pecking order so they know how to seek the appropriate help to complete a task.

ISO 9001 AND ISO 14001

- Every engineering company or organization has its own quality management system and continuous process improvement goals that cover all segments of a company or organization.
- Examples of an engineering department's processes that are covered by ISO 9001 and by ISO 14001 include document control procedures; change order procedures; training procedures; laboratory equipment calibration procedures; laboratory equipment precision and accuracy determination procedures; error control and elimination procedures; lessons-learned procedures; customer communication procedures; in-house (domestic and international) communication procedures; regulatory requirement procedures; product design, development, review, verification, and validation procedures; and nonconforming product control procedures, among others.
- All process procedures and their records must be documented and accessible to all subordinates.
- All subordinates should be familiar with ISO 9001 and ISO 14001 quality and environmental management procedures and process improvement goals.
- The engineering manager must adhere closely to the company's quality control program—namely, ISO 9001—and work to make continual process improvements.
- Subordinates should be encouraged to provide input regarding process improvement that does not deviate from the company's quality control procedures.
- The same rules apply to the company's environmental management system and ISO 14001; subordinates should become involved in climate change mitigation, reducing energy and water consumption, waste management, etc.

10 Engineering Team Building

SUBGROUPS IN AN ENGINEERING DEPARTMENT

An engineering department is normally composed of several subgroups depending on how diversified the engineering knowledge must be. These engineering subgroups can vary from electrical circuit design to mechanical tool design to materials. Each subgroup can be composed of one or several engineers and technicians. Subgroups that have more than one subordinate must have an assigned leader, who reports to the engineering department manager. Subgroup leaders continue to perform tasks in their field of expertise but are also responsible for the performance of other personnel in the subgroup. Engineering department managers distribute assignments to subgroups through their subgroup leaders. In a large engineering department (say, more than 10 subordinates), it becomes very difficult for an engineering manager to track all of the details of every task. Such details are supervised by the subgroup leader, who also works to resolve any problems. The engineering department manager gathers weekly summary information about tasks from subgroup leaders.

As an example, consider an engineering department for a computer component design and manufacturing company. It has 40 engineers and technicians divided among several subgroups, such as component design, manufacturing, tool design, automation, component testing, and customer support. Most assignments for these subgroups are multidisciplinary. If correlation between a customer's tester and a company's tester is required, both a customer test engineer and a customer support engineer will be assigned to the task. If a new computer component assembly production line has to be set up, a manufacturing engineer, a couple of tool designers, and an automation engineer will be assigned to the task.

In another example, an engineering department for a company producing offshore oil platform load movement systems has 15 engineers and technicians, who are divided among such subgroups as structural design, system control design, hydraulics design, tool design, customer support, and project management. Again, most assignments for these subgroups are multidisciplinary. The installation of new load movement equipment and training the customer's personnel in using the new system require a customer support engineer, a system control design engineer, and a technician.

An engineering department that develops internal combustion engines for a large automotive company might have 22 engineers and technicians. The subgroups for this engineering department include materials, heat transfer and fluid mechanics, component and assembly design, prototype construction, and testing. Almost all of the subgroups interact during the development phase of an engine, when the fluid mechanics engineer might work side by side with a materials engineer, a component and assembly designer, and a prototype construction and test engineer in order to determine the working characteristics of the new engine cooling system.

In today's global environment, subgroups in an engineering department can include global consultants, global freelancers, and engineers or technicians from the company's international divisions. These global subgroups can be much more difficult to lead and can require an excessive time commitment from subgroup leaders, which can degrade the subgroup's performance and raise the leader's stress level to unacceptable levels. Engineering managers should be careful to limit the leadership responsibilities of subgroup leaders to no more than 10 to 15 hours a week. Larger subgroups can be broken down into two or three sub-subgroups to more appropriately balance the leaders' responsibilities; for example, a materials subgroup can be divided into two sub-subgroups, one responsible for plastics and the other for metals.

Engineering managers must choose subgroup leaders very carefully. Subordinates must be groomed to be subgroup leaders. The engineering manager should determine if a person has the necessary leadership capabilities. Is this person an extrovert? Does this person have the desire and knowledge to help others in a particular subgroup? Does this person have the planning ability and ingenuity to get tasks done? Is this person persistent enough to get tasks completed on time and with the highest quality in spite of difficulties and hurdles? Can this person be depended on to solve technical and nontechnical issues? Does this person like to brainstorm to solve technical and nontechnical problems? Does this person listen to the input of others? Does this person tend to consider all aspects of a problem carefully rather than shooting from the hip? Is this person ready to grow within the company rather than jumping ship? Does this person have rivals in the group who would prefer to have the leadership spot?

When a new subgroup leader has been chosen, the engineering manager should discuss future plans within the organization for that person. Some potential subgroup leaders may not want the extra burden of managing others. They may prefer to focus on their field of expertise and not be distracted by other responsibilities. To provide incentive, the engineering manager should discuss the financial advantages of being promoted to a leadership role, as well as possible future assignments.

Engineering managers should meet at least once a week with their subgroup leaders to review the status of tasks within the various subgroups. Assigning new tasks to subordinates should be done by the subgroup's leader. Engineering managers should be careful not to interfere with the performance of a task in a subgroup without the knowledge of the subgroup leader. Engineering managers should have at least biweekly, one-on-one meetings with subgroup members to hear first-hand how the tasks are proceeding rather than depending solely on the input of subgroup leaders. Engineering managers must be aware of everything going on in their department to avoid unexpected surprises and constant firefighting.

Another type of team in an engineering department is formed specifically for a particular project, such as a development project, process improvement project, or new manufacturing line qualification project. These teams have specific goals and timelines to complete a project. More than likely, people from other departments might also join these project teams. In the international arena, a large project team can grow more diversified and have two or three team leaders (e.g., one in each country). All of these team leaders would report to the engineering manager. Task coordination and continuous communication are crucial for the success of international project teams. The same considerations apply when choosing project team leaders as for choosing subgroup leaders, but project team leaders must be more diversified in engineering knowledge and more resourceful in finding different ways to accomplish a project's tasks. Several subordinates can be assigned to a project team, and a subordinate can be a member of several project teams. A subordinate can serve on a project team for a couple of months or a couple of years. Such subordinates report first to their project leaders or managers. When reviewing a subordinate, the engineering manager should seek input from the person's project leader or manager.

INTERNATIONAL TEAMS

When an engineering department has international subgroups, it is advisable to assign leaders for such subgroups from the particular country in which they are located. Work habits and traditions vary from country to country, and a local subgroup leader can achieve harmony within the group with greater ease. Members of a subgroup in Japan, for example, that provides customer support should be brought to the United States periodically for training to keep them up to date on advancing technologies in the field. Whenever possible, the engineering manager should give reviews to subordinates in a foreign subgroup in that particular country. All salary increases and bonuses should be commensurate with that country's standards. All employee hiring and firing should be according to the country's ground rules and regulations.

Suppose a subgroup in Malaysia is responsible for developing automation equipment for the company's Malaysian production lines. This subgroup has engineers and technicians from both the United States and Malaysia. The subgroup leader should be Malaysian rather than American, because a Malaysian subgroup leader will be more effective at coordinating and getting tasks done. Subgroup members from the United States should primarily carry out training and consulting functions. When training must be carried out in the language of the foreign country, training documents and materials must be translated to that language. In Malaysia, for example, the training documents and materials might have to be translated into Malay.

Teams in a foreign country should learn about the history of that country and the local customs and traditions. It is extremely helpful if subordinates can learn to speak some useful phrases in the native language, such as "good morning," "good evening," "please," "thank you," "how are you," and "excuse me." Also, engineering managers and subordinates should try to change any less desirable work habits they might encounter in a foreign country (e.g., lackadaisical meetings, long midday siestas).

MULTIDISCIPLINARY TEAMS

Multidisciplinary teams are a fact of life in modern engineering departments. Some subordinates might be assigned to teams in other departments, and some people from other departments might join teams in the engineering department. A photolithography engineer might be on a receiving and inspection team responsible for determining the quality of incoming photoresist. Or, a purchasing agent who has extensive experience in sourcing special steels can be on an offshore oil platform equipment design team. These types of multidisciplinary teams broaden the subordinates' knowledge base, teach resourcefulness, and demonstrate how to achieve desired results in a complex environment. Membership in such teams can help subordinates learn how to interact better with customers, suppliers, salespeople, purchasing agents, managers, subgroup leaders from other departments, legal representatives, etc.

Imagine a multidisciplinary team in South Korea that includes two customer engineers and two subcontractor engineers. In such a multinational environment, the tasks become more exciting and complex. Such a multidisciplinary, international team environment requires clear task instructions and doable task completion dates. It is advisable for team leaders to get verbal task commitments in writing. Meeting minutes should always be published, document controlled, and distributed to all parties involved with the team's project. In a multidisciplinary team environment, relentless persistence to complete a task might work against the team. Team members must learn to be patient and not to antagonize others when trying to complete a task on time. They must also learn how and when to apply the right amount of pressure on others to achieve what they are after.

Behavior in a multidisciplinary or international meeting environment must be appropriate. Effective team leaders require extensive training in the multidisciplinary, international arena. Consider engineers and a customer visiting a supplier in Japan, for example. They must respect and follow the normal meeting etiquette of the Japanese supplier. This particular engineering team includes a senior design engineer from the United States and an English-speaking senior quality engineer from a South Korean subsidiary. The customer decides to send along a senior purchasing agent from their Singapore office. The Japanese supplier will bring a team of ten people to the meeting, including their president, senior salespeople, senior manufacturing engineers, senior quality engineers, and a translator. The engineering team and customer are there to negotiate technical specifications for a critical computer component and the cost of the component, which the Japanese supplier will provide for volume production lines in the South Korea facility. Also discussed will be a potential production ramp-up plan. The meeting is planned to run for two days, from 9 a.m. to 9 p.m. each day. The engineers and customer are guests of the Japanese supplier, and the Japanese supplier is the meeting organizer. The engineering team and customer must sit where they are shown in the conference room—most likely at the far section of the conference table, away from the entry door to the room. The engineering team and customer can request items for discussion to be added to the meeting agenda. Any mementos offered by the Japanese subcontractor should be graciously accepted, as should dinner invitations extended by the Japanese subcontractor, even if the team members are very tired due to the long meeting and jet lag. Because all communication is conducted in both English and Japanese, everything being said must be translated simultaneously. All meeting minutes and action items must be signed and released in both English and Japanese.

Subordinates must be taught how to participate in such multidisciplinary, international meetings before being allowed to enter the lion's den by themselves. Engineering managers should bring subordinates with them to these types of meetings so they can observe and learn. Subordinates must study a particular country's business language, business ethics, history, customs, and traditions before being invited to attend such meetings.

TEAM LEADERS

Team leaders can be subgroup leaders, project leaders, or leaders specifically assigned for a multidisciplinary, international meeting. Subgroup leaders must be extroverts who willingly accept the additional leadership responsibilities. They do not have to be the most technologically advanced person in the particular subgroup. Subgroup leaders should avoid creating friction or fostering jealousy among subgroup members. They should enjoy helping members of the subgroup successfully complete the tasks assigned, and they should know when to seek the help of the engineering manager. Subgroup leaders should also have a knack for dealing with other departments and managers in the company.

Not every subgroup leader can be expected to be an effective project leader in a multidisciplinary, international environment, where team leadership becomes more complicated and stressful. Successful project leaders in a multidisciplinary, international environment have extraordinary communication skills and discipline. They are able to oversee all communications related to the project, in addition to preparing schedules and cost performance indices for projects and presenting them periodically to upper management. They are good listeners, team motivators, team advocates, and team ambassadors. They know how to be the project drivers in an international arena. They are able to filter and present information (e.g., specifications, standards, scope or schedule changes) clearly to their team members. Successful project leaders gain the respect of their domestic and international team members, as well as that of customers, subcontractors, and regulatory agencies.

The engineering manager must groom subordinates to be effective project leaders in a multidisciplinary, international environment. If project leadership potential becomes apparent in a subordinate, the engineering manager can assign that person a small, clearly defined internal project and project team. If that project is completed successfully, then the novice project leader can be assigned multidisciplinary but still internal projects. If these projects are completed with flying colors, then that person can move on to leading a domestic customer's project to learn how to deal with customer project managers, subcontractors, and regulatory agencies. The next step for the successful project leader would be taking on projects in a multidisciplinary, international environment. Some project leaders will not want to deal with the challenges presented by such projects. Some may not like to deal with people of different cultures or do not like to travel internationally, especially for an extended length of time.

Not every subgroup leader can lead a multidisciplinary, international team. Not everyone has the broad perspectives and in-depth knowledge of the company and its customers required, or the people skills necessary to conduct a multidisciplinary, international meeting. An engineering manager who sees multidisciplinary, international team leadership potential in a subordinate should allow that person to attend international meetings to observe and learn. The engineering manager can give the subordinate in-depth information about the company, its customers, and its subcontractors, especially with regard to the technology developed by them, their business structure, and future outlook. After receiving extensive grooming and further enhancement of the people skills necessary to participate in multidisciplinary, international meetings, this person can become a valued team leader for such meetings.

CROSS-TRAINING

Engineering managers can prepare subordinates for leadership positions by cross-training them in different departments of the company, international sites of the company, customers, and subcontractors to expose them to various types of technology and business fields. Such training should be mutually agreed-upon, planned well ahead of time, and documented in the subordinate's reviews. In the author's experience, this type of multidisciplinary, international training can take 2 to 4 years, depending on the resourcefulness and talents of the person being trained, particularly that person's ability to handle the stress that can come with leadership roles.

CHECKLIST FOR CHAPTER 10

SUBGROUPS IN AN ENGINEERING DEPARTMENT

- An engineering department is normally composed of several subgroups depending on how diversified the engineering knowledge must be.
- Engineering subgroups can vary from electrical circuit design to mechanical tool design to materials.
- Subgroups that have more than one subordinate must have an assigned leader, who reports to the engineering department manager.
- Engineering department managers distribute assignments to subgroups through their subgroup leaders.
- In a large engineering department, it becomes very difficult for an engineering manager to track all of the details of every task. Such details are supervised by the subgroup leader, who also works to resolve any problems.
- The engineering department manager gathers weekly summary information about tasks from subgroup leaders.
- Subgroups should work harmoniously together to complete multidisciplinary tasks.
- Engineering managers should be careful to limit the leadership responsibilities of subgroup leaders to no more than 10 to 15 hours a week.
- Engineering managers must choose subgroup leaders very carefully.
- The engineering manager should determine if a person has the necessary leadership capabilities.
- Is this person an extrovert?
- Does this person have the desire and knowledge to help others in a particular subgroup?
- Does this person have the planning ability and ingenuity to get tasks done?
- Is this person persistent enough to get tasks completed on time and with the highest quality in spite of difficulties and hurdles?

- Can this person be depended on to solve technical and nontechnical issues?
- Does this person like to brainstorm to solve technical and nontechnical problems?
- Does this person listen to the input of others?
- Does this person tend to consider all aspects of a problem carefully rather than shooting from the hip?
- Is this person ready to grow within the company rather than jumping ship?
- Does this person have rivals in the group who would prefer to have the leadership spot?
- When a new subgroup leader has been chosen, the engineering manager should discuss future plans within the organization for that person.
- To provide incentive, the engineering manager should discuss the financial advantages of being promoted to a leadership role, as well as possible future assignments.
- Assigning new tasks to subordinates should be done by the subgroup's leader. Engineering managers should be careful not to interfere with the performance of a task in a subgroup without the knowledge of the subgroup leader.
- Engineering managers should have at least biweekly, one-on-one meetings with subgroup members to hear first-hand how the tasks are proceeding rather than depending solely on the input of subgroup leaders.
- When reviewing a subordinate on a multidisciplinary, international team, the engineering manager should seek input from the person's project leader or manager.

INTERNATIONAL TEAMS

- When an engineering department has international subgroups, it is advisable to assign leaders for such subgroups from the particular country in which they are located.
- Members of a subgroup in a foreign country should be brought to the United States periodically for training to keep them up to date on advancing technologies in the field.
- Whenever possible, the engineering manager should give reviews to subordinates in a foreign subgroup in that particular country.
- All salary increases and bonuses should be commensurate with that country's standards.
- All employee hiring and firing should be according to the country's ground rules and regulations.
- Subgroup members from the United States should primarily carry out training and consulting functions.
- Teams in a foreign country should learn about the history of that country and the local customs and traditions.
- Engineering managers and subordinates should try to change any less desirable work habits they might encounter in a foreign country.

MULTIDISCIPLINARY TEAMS

- A multidisciplinary, international team environment requires clear task instructions and doable task completion dates.
- It is advisable for team leaders to get verbal task commitments in writing.
- Meeting minutes should always be published, document controlled, and distributed to all parties involved with the team's project.
- In a multidisciplinary team environment, relentless persistence to complete a task might work against the team. Team members must learn to be patient and not to antagonize others when trying to complete a task on time.
- Team members must learn how and when to apply the right amount of pressure on others to achieve what they are after.
- Effective team leaders require extensive training in the multidisciplinary, international arena.
- Subordinates must be taught how to participate in such multidisciplinary, international meetings before being allowed to enter the lion's den by themselves.
- Engineering managers should bring subordinates with them to these types of meetings so they can observe and learn.
- Subordinates must study a particular country's business language, business ethics, history, customs, and traditions before being invited to attend such meetings.

TEAM LEADERS

- Subgroup leaders must be extroverts who willingly accept the additional leadership responsibilities.
- Subgroup leaders should avoid creating friction or fostering jealousy among subgroup members.
- Subgroup leaders should enjoy helping members of the subgroup successfully complete the tasks assigned, and they should know when to seek the help of the engineering manager.
- Subgroup leaders should have a knack for dealing with other departments and managers in the company.
- Successful project leaders in a multidisciplinary, international environment have extraordinary communication skills and discipline.
- Project leaders oversee all communications related to a project, in addition to preparing schedules and cost performance indices for projects and presenting them periodically to upper management.
- Project leaders are good listeners, team motivators, team advocates, and team ambassadors.
- Project leaders know how to be the project drivers in an international arena.
- Project leaders are able to filter and present information (e.g., specifications, standards, scope or schedule changes) clearly to their team members.

- Successful project leaders gain the respect of their domestic and international team members, as well as that of customers, subcontractors, and regulatory agencies.
- If project leadership potential becomes apparent in a subordinate, the engineering manager can assign that person a small, clearly defined internal project and project team. Potential project leaders will continue to grow as they take on more complicated projects in a multidisciplinary, international environment.
- Not every subgroup leader can lead a multidisciplinary, international team. Not everyone has the broad perspectives and in-depth knowledge of the company and its customers required, or the people skills necessary to conduct a multidisciplinary, international meeting.
- Project leaders in a foreign country should learn about the history of that country and the local customs and traditions.
- An engineering manager who sees multidisciplinary, international team leadership potential in a subordinate should allow that person to attend international meetings to observe and learn.
- The engineering manager can give project leaders in-depth information about the company, its customers, and its subcontractors, especially with regard to the technology developed by them, their business structure, and future outlook.

CROSS-TRAINING

- Engineering managers can prepare subordinates for leadership positions by cross-training them in different departments of the company, international sites of the company, customers, and subcontractors to expose them to various types of technology and business fields.
- Such training should be mutually agreed-upon, planned well ahead of time, and documented in the subordinate's reviews.

11 Upper Management, Customer, Subcontractor, and Regulatory Relationships

UPPER MANAGEMENT RELATIONSHIPS

Engineering departments are the foundation of technology companies. Engineering managers and their subordinates must interact on a daily basis with people in other departments and with upper management. Engineering managers must train subordinates how to deal with ill-timed interruptions, such as a sales executive who constantly pesters the engineering manager's subordinates about a pet project. If such interruptions become excessive, especially by an upper manager, subordinates should report the situation to their managers, who can deal with the situation by offering to provide daily updates themselves. Upper managers can be politely asked not to interfere with the work of the engineering manager's subordinates. It is the engineering manager's responsibility to protect subordinates from upper management pressure. In some cases, it might become necessary for engineers to work from home or in a library so that they can work undisturbed and complete their tasks on time. Colleagues and people from other departments can also be a source of disruptions. Ideally, upper managers and people from other departments should first approach the engineering manager for task requests or to get updates on the status of tasks.

On occasion, subordinates might encounter some problems getting work done in other departments. For example, a subordinate might find it necessary to frequently nudge purchasing about an urgently needed item, but such persistence could cause some friction with the purchasing agent. Subordinates should be trained to come to their manager to ask for help when necessary. In this case, the manager and subordinate together could ask the purchasing agent to provide daily updates on delivery of the item. If such an approach proves ineffective, then the purchasing manager could be brought into the picture.

Engineering managers must regularly update their supervisors and other upper managers regarding the status of projects in their department. Engineering managers must be concise and clear when summarizing the status of projects. When it is necessary for the engineering manager and a subordinate to give a status presentation at an upper management meeting, both must be well prepared by practicing the presentation several times beforehand. Also, there should be no errors in presentation materials that will be handed out in the meeting. Any presentation should include both good news and bad, such as tasks that are on the critical path of a project but are delayed. The engineering manager should be prepared to present options to remedy the situation, but upper management might be able to contribute several good ideas.

A senior engineer and I had to make a technical presentation to the board of directors of the company to obtain funding for a large production line automation project that was necessary to accommodate the tighter tolerances required by the products. After preparing a feasibility study for the automation project, we presented our findings to my subordinates and supervisor. They provided some good input, and we updated our feasibility study accordingly. We were then assigned a half-hour slot to give our presentation to the board. Our 20-page handout summarized the feasibility study for the automation project. After the presentation, the board had to decide whether or not to fund the automation project. The board members asked several questions during the presentation, but none of the questions addressed any technical aspects of the project. Everyone was focused on how to reduce the projected cost of the automation and on the company's return on investment. The board members gave us some good suggestions and asked us to reevaluate the project for various options, such as for different project sites or different automation system maintenance teams. We were given a month to perform new feasibility studies for the various options and to prepare a presentation of our results to the board. Although we carried out several different feasibility studies, the original proposal always came out on top. After the second presentation, the board of directors approved funds for the automation project. The keys to our success were that my senior engineer and I prepared the feasibility studies meticulously and in utmost detail, practiced several times beforehand, and made sure that there were no errors in the presentation materials handed out in the board meeting.

Engineering managers and their subordinates should not make commitments on behalf of people outside of the engineering department. A task requiring the involvement of people from other departments should always be discussed with them and their supervisors before the scope of the task and its completion date are decided upon. All task agreements should be backed up in writing and provided to the persons involved in other departments and their supervisors. Even within the engineering department, all potential tasks should be discussed with the people who will be involved before proceeding. Engineering managers and their subordinates should always remember that clear and concise communication and strict documentation are important for success in a multidisciplinary work environment.

If an issue cannot be resolved through normal negotiations with other department heads and is hampering the work of subordinates, the engineering manager should discuss the issue with his or her supervisor. Going all the way up the ladder to the company president to solve the problem should be a last resort. The engineering manager cannot rock the boat for every little issue, but seeking the help of higher ups might become necessary for significant issues that negatively affect the working atmosphere and progress of work in the engineering department.

A good example of this is an information technology (IT) manager I once worked with. He was always late in his commitments to upgrade the engineering department's software and computer equipment. He usually pleaded the need to stay within his budget and delayed new purchases as much as possible. I had to constantly keep after him to update the software and computer equipment in the engineering department. I finally went to my supervisor, and we had a meeting with the IT manager about this issue, but nothing changed. My subordinates complained constantly about being behind in the necessary software and computer technologies. After discussing the problem with other department managers, I discovered that they were having similar problems with the IT manager, but it was not as significant a problem for them. Finally, with the permission of my supervisor, I called a meeting for all department heads, the president of our company, and the IT manager. I prepared a presentation that summarized the problems we were having as a result of the outdated software and computer equipment, such as software incompatibility with our customers and subcontractors and slow calculation times. After my presentation, the president ordered the IT manager to bring all software and computer equipment up to date within 2 weeks and to not delay any future software and computer equipment updates. The president also asked the IT manager to modify his budget to include the necessary upgrades. After much frustration, I was finally able to solve the IT issue by enlisting the help of my supervisor, other department heads, and the company president.

Engineering managers and their subordinates should always praise people outside of the engineering department for a job well done. The supervisors of the people who performed well on a job should always be copied in on any written praise. Maintaining good rapport with personnel in other departments can go a long way toward successful completion of an engineering department's tasks. Such departments would include purchasing, information technology, sales and marketing, finance, receiving and inspection, legal, human resources, and manufacturing. One never knows whose help might be needed for a crucial project.

Sometimes the rumor mill will run rampant because of an unfortunate situation arising in the engineering department, in other departments, in upper management, or even throughout the company. It could be a major shakeup in one of the departments or in upper management. The president of the company might have been fired by the board of directors. Another company might be trying to buy out the company, or *vice versa*. Perhaps quarterly earnings were unexpectedly low and the company's stock prices have taken a hit. There might be a companywide layoff on the horizon. The engineering manager should hold a department meeting as soon as possible to share what he or she knows about the situation and to put a stop to the rumors. The engineering manager must not make assumptions and should not exaggerate any details. If necessary, the engineering manager's supervisor or other upper management person such as the human resources manager could be invited to the meeting to help explain what is going on.

CUSTOMER RELATIONSHIPS

Subordinates must be trained how to deal with every level of a customer's organization and how to gain a customer's trust. Subordinates should document every communication with a customer and copy their manager and other relevant people in the company. Subordinates should not make commitments on behalf of others in the company. If an important issue arises that they cannot address properly, they should seek their manager's help. Also, subordinates should never deal directly with the upper management of a customer. A subordinate who encounters difficulties responding to an inquiry by a customer should not try to provide an answer right away. The subordinate should consult with the engineering manager or other appropriate people in the company before getting back to the customer.

Customer meetings, either domestic or international, require careful preparation and attention to such details as drawing up the meeting agenda, printing the meeting handouts, determining what topics should be avoided during the meeting, protecting the intellectual property of other customers, having translators available, placing flags on the conference table for the various nations involved, arranging for catering, assigning someone to take the meeting minutes, distribution of the meeting minutes, etc. Engineering managers and their subordinates should be aware that most customer meetings will run over the allocated time and should be prepared for that to happen. Subordinates cannot leave a customer meeting to go pick up their children from school. Subordinates who have appointments or other commitments scheduled on a customer meeting day should reschedule them to another day. If it becomes necessary for subordinates to cut short their vacation, they should be well rewarded for their sacrifice, such as with an all-expenses-paid vacation to the same location at another time.

It is not unusual to encounter frequent personnel changes when dealing with a customer. The customer's project manager who is serving as the liaison between the engineering group and the customer's organization might change. The customer's purchasing agent might change. A customer who insists on having one of their own employees observing the engineering department's project team during the execution of a particularly important project might then switch out that observer in the middle of the project. The customer might even gain a new president who is not on good terms with the engineering firm's president. Engineering managers and their subordinates must learn how to deal effectively with such customer personnel changes. It will usually be necessary to bring the customer's new person up to speed with regard to the status of the project. In such cases, it is very helpful to have a completely documented history of the project, its specifications, and any change orders.

When someone from a customer's upper management contacts an engineering manager's subordinate to ask a question or request something, the subordinate should immediately get the engineering manager involved. Subordinates should not make any commitments to a customer, domestic or international, on behalf of the department or company. Communication between the engineering manager's company and the customer should take place at equivalent levels; that is, the company's president should talk to the customer's president, the company's sales executive should talk to the customer's purchasing executive, etc. Especially in the international arena, the pecking order for communications and commitments is very important.

On one occasion, a U.S. customer assigned a full-time resident engineer to observe the production of a computer component on my company's specially designed mass production lines in South Korea. The resident engineer's primary function was quality control for the state-of-the-art component being built for his company. He was also observing how our company was protecting his company's intellectual property. Essentially, I treated the customer's resident engineer as my own subordinate, and his manager asked for my input for his employee's annual review. My team and I assisted him as much as possible during his tenure, even stepping in to help out when he became ill with food poisoning. He stayed two years in South Korea. My customer's executives and the resident engineer were very appreciative of our treatment of him during his assignment in our production facility.

Engineering managers and their subordinates must listen carefully to customers and support them in any way possible. My company built a computer component to a customer's specifications and sent a shipment of the component to the customer. The component met all of the external dimensions specified by the customer, but the customer's width specification was in error. The customer's somewhat panicked project manager called and pleaded for 100% sorting of the computer component for the width dimension before shipment. We immediately went to work to come up with a go/no-go fixture to inspect the width dimension of every computer component before shipment. We lost about 8% of the production of the component, but the customer paid for the rejected products. We shipped product with the updated width dimension, and every component fit into the assembly without any issues. The ecstatic customer praised my team's exceptional and timely efforts and wrote commendation letters to our upper management, including the company's president. The commendation letters resulted in higher raises and bonuses that year for everyone on the team, and the customer continued to buy our products for many years to come.

An important aspect of customer relationships is protecting each customer's intellectual property. A company's customers might include several who are competitors in a very dynamic market. An impermeable knowledge wall must be maintained for each customer's product specifications and design characteristics. If two competing customers happen to have meetings scheduled at the company at the same time, they must not be allowed to run into each other. If a customer wants their design drawings to be kept in a vault and asks that no copies be made, those requests must be respected. Access to a customer's confidential drawings should be restricted to only those subordinates who have been approved to sign them out. Every confidential drawing and specification that has been checked out must be returned to the vault each night. If a customer asks that their product's production line be separate from other production lines in the facility, only the people involved with that customer's product should be allowed to enter that specific area of production. In some particularly sensitive cases, it might be necessary to isolate a specific production line with plastic walls to limit access.

SUBCONTRACTOR RELATIONSHIPS

A subcontractor can make or break a business. How subcontractors are chosen and how they are monitored are the two most important aspects of dealing with subcontractors. Engineering managers and their subordinates more than likely will get involved in qualifying subcontractors, at least in the technical arena. Qualification of a subcontractor should be a very thorough process that examines the subcontractor's personnel and process control verification. A subcontractor's committed task completion date can be compromised by, for example, the loss of critical personnel, a strike, cash flow issues, or a change in job priority. A sole-source subcontractor might decide to shut down operations right in the middle of an important project. The company's representative at a subcontractor's site should immediately report any unusual issues, such as a slowdown in the project or quality problems. Overseeing a subcontractor onsite is crucial, especially in foreign countries. For especially important projects, a reliable subordinate should be in place at the subcontractor's site on a full-time basis.

In certain cases, it might become necessary to provide technical, personnel, or financial support to a subcontractor to help them complete the job. It is not unusual to have to pay subcontractors upfront so they can cover payroll or equipment costs associated with a project. When a sole-source subcontractor in Japan was experiencing terrible process yields, a senior manufacturing engineer and I went to Japan to help sort out their process issues and bring their yields up to acceptable levels.

The engineering team should listen carefully to their subcontractors, who might have more expertise in a particular field; for example, input from a subcontractor led to a change in the design of an offshore oil platform that allowed for special steel casting for strength and shock resistance. In another case, a customer had very tight tolerances for the bronze coating of steel trolley wheels. Our highly regarded bronze coating subcontractor believed that the tight tolerances were not practical at the corners of the wheels and could not be done. My subcontractor and I had to convince the customer to loosen the bronze coating tolerances to a reasonable level so the project could proceed. Another time, my subordinates and I had to completely depend on our subcontractor's input on surface contamination. We sent our contaminated computer components to the subcontractor, who performed auger spectroscopy analysis on every component and reported the results. We depended totally on his contamination reports to properly maintain our cleanrooms. In yet another case, an assembler on an engine prototype line at our subcontractor found an error in our engine design. Whenever we tested our prototype engines, we encountered high oil pressures. One day, a line assembler pulled me aside and told me that in all of the other prototype engines he had built the oil pressure relief valve exit holes were larger. As a result of his valuable input, the high oil pressure problem was solved by redesigning the undersized oil pressure relief valve exit hole.

REGULATORY RELATIONSHIPS

Engineering managers and their subordinates will deal with numerous regulatory and certification agencies and companies during the life of a project. If an offshore oil rig equipment is being designed and built for use in Norway, Det Norske Veritas (DNV) will be involved in the design and test certification of the product. If a similar product is intended for use in the United States, the American Bureau of Shipping (ABS) will be involved in the design and test certification of the product. If a product to be marketed in Europe must be certified for safety (e.g., a product designed for a Class 1, Division 1 hazardous location), then European Conformity (CE) certification must be obtained for that product. If the same product is to be marketed in the United States, Underwriters Laboratories (UL) certification for safety should be obtained. Customers will require material certificates for every material lot being used in their products, so it is necessary to know how to deal with material certification laboratories to obtain material certificates in a timely fashion. If precise and accurate measurements on the production line are required, it will be necessary to obtain certified standards from standards certification companies for length, weight, temperature, time, etc. Manufacturers of equipment that will be shipped to Russia must apply for and receive GOST-R certificates of conformity for customs clearance before shipment. Engineering managers and their subordinates should deal with an expert agency in the particular country for which they are seeking certification—such as Russia's GOST-R certificates of conformity for customs clearance. Otherwise, they almost certainly will be wasting their time trying to obtain certification on their own.

Engineering managers and their subordinates must make sure that a regulatory agency's certificate is complete and signed appropriately. I encountered a problem with customs in South Korea when several spare parts included on the final packing list from an inspector's certification report were missing. I had overlooked this unfortunate error in the inspector's report, as did the shipping department, the forwarding agency, and the bank that paid my company the milestone payment. The customer in South Korea could not clear the shipment from customs, and I had to scramble to get a corrected certification report to South Korea so the shipment could clear customs.

In another example, certification testing of our offshore oil rig equipment was expected to take a week, including functional tests, load tests, nondestructive tests, mechanical tests, electrical tests, and environmental tests. Two inspectors from the regulatory agency were scheduled to observe and certify all of the testing. The engineering team pretested the equipment to make sure that all regulatory agency requirements were met before the inspectors arrived to our facility, but somehow we overlooked the traceability of materials used in some steel parts to their origins. The regulatory agency and customer both required complete traceability of raw materials for steel plates, forgings, castings, fasteners, pins and shafts from their original heat lots to finished products, and it was necessary to laser engrave the heat lot number and material vendor's designation on each finished part. I had told my manufacturing engineering lead about this traceability requirement at the beginning of the project, but this information did not trickle down to some of the machinists. The certification tests had to be delayed for a week so we could address this oversight. Fortunately, we were able to determine the missing heat lot and vendor information with the help of stockroom personnel and records.

When product failures occur during certification tests, the certification inspectors should be sent home. They should not be allowed to sit around while everyone is scrambling to fix the problem. It is better to fix the problem while certification inspectors are not looking over your shoulder and to reschedule the tests after successful retesting.

It is helpful to become familiar with the quirks of various certification inspectors encountered through the years. One certification inspector I knew for many years always made us check the tightening torque for every bolt on a piece of equipment. It was important to him that we met all the bolt tightening torque specifications. Whenever he showed up for an equipment certification, we were always ready for him with a calibrated torque meter.

CHECKLIST FOR CHAPTER 11

Upper Management Relationships

- Engineering managers must train subordinates how to deal with ill-timed interruptions, such as a sales executive who constantly pesters the engineering manager's subordinates about a pet project.
- If interruptions become excessive, especially by an upper manager, subordinates should report the situation to their managers, who can deal with the situation by offering to provide daily updates themselves.
- Upper managers can be politely asked not to interfere with the work of the engineering manager's subordinates.
- It is the engineering manager's responsibility to protect subordinates from upper management pressure.
- On occasion, subordinates might encounter some problems getting work done in other departments. For example, a subordinate might find it necessary to frequently nudge purchasing about an urgently needed item, but such persistence could cause some friction with the purchasing agent. Subordinates should be trained to come to their manager to ask for help when necessary.
- Engineering managers must regularly update their supervisors and other upper managers regarding the status of projects in their department.
- Engineering managers must be concise and clear when summarizing the status of projects.
- When it is necessary for the engineering manager and a subordinate to give a status presentation at an upper management meeting, both must be well prepared by practicing the presentation several times beforehand.
- Any presentation should include both good news and bad, such as tasks that are on the critical path of a project but are delayed.
- Engineering managers should be prepared to present options to remedy the bad news.
- Engineering managers and their subordinates should not make commitments on behalf of people outside of the engineering department.
- A task requiring the involvement of people from other departments should always be discussed with them and their supervisors before the scope of the task and its completion date are decided upon.
- All task agreements should be backed up in writing and provided to the persons involved in other departments and their supervisors.
- Even within the engineering department, all potential tasks should be discussed with the people who will be involved before proceeding.
- Engineering managers and their subordinates should always remember that clear and concise communication and strict documentation are important for success in a multidisciplinary work environment.
- If an issue cannot be resolved through normal negotiations with other department heads and is hampering the work of subordinates, the engineering manager should discuss the issue with his or her supervisor. Going all the way up the ladder to the company president to solve the problem should be a last resort.

- The engineering manager cannot rock the boat for every little issue, but seeking the help of higher ups might become necessary for significant issues that negatively affect the working atmosphere and progress of work in the engineering department.
- Engineering managers and their subordinates should always praise people outside of the engineering department for a job well done.
- The supervisors of the people who performed well on a job should always be copied in on any written praise.
- Maintaining a good rapport with personnel in other departments can go a long way toward successful completion of an engineering department's tasks.
- When rumors are circulating, the engineering manager should hold a department meeting as soon as possible to share what he or she knows about the situation and to put a stop to the rumors.
- If necessary, the engineering manager's supervisor or other upper management person such as the human resources manager could be invited to the meeting to help explain what is going on.

CUSTOMER RELATIONSHIPS

- Subordinates must be trained how to deal with every level of a customer's organization and how to gain a customer's trust.
- Subordinates should document every communication with a customer and copy their manager and other relevant people in the company.
- Subordinates should not make commitments on behalf of others in the company.
- If an important issue arises that subordinates cannot address properly, they should seek their manager's help.
- Subordinates should never deal directly with the upper management of a customer.
- A subordinate who encounters difficulties responding to an inquiry by a customer should not try to provide an answer right away. The subordinate should consult with the engineering manager or other appropriate people in the company before getting back to the customer.
- Customer meetings, either domestic or international, require careful preparation and attention to details.
- Engineering managers and their subordinates should be aware that most customer meetings will run over the allocated time and should be prepared for that to happen.
- It is not unusual to encounter frequent personnel changes when dealing with a customer.
- Engineering managers and their subordinates must learn how to deal effectively with such customer personnel changes.
- It will usually be necessary to bring the customer's new person up to speed with regard to the status of the project. In such cases, it is very helpful to have a completely documented history of the project, its specifications, and any change orders.

- When someone from a customer's upper management contacts an engineering manager's subordinate to ask a question or request something, the subordinate should immediately get the engineering manager involved.
- Subordinates should not make any commitments to a customer, domestic or international, on behalf of the department or company.
- Communication between the engineering manager's company and the customer should take place at equivalent levels. Especially in the international arena, the pecking order for communications and commitments is very important.
- Engineering managers and their subordinates must listen carefully to customers and support them in any way possible.
- An important aspect of customer relationships is protecting each customer's intellectual property.
- An impermeable knowledge wall must be maintained for each customer's product specifications and design characteristics.
- If two competing customers happen to have meetings scheduled at the company at the same time, they must not be allowed to run into each other.
- If a customer wants their design drawings to be kept in a vault and asks that no copies be made, those requests must be respected.
- If a customer asks that their product's production line be separate from other production lines in the facility, only the people involved with that customer's product should be allowed to enter that specific area of production. In some particularly sensitive cases, it might be necessary to isolate a specific production line with plastic walls to limit access.

Subcontractor Relationships

- How subcontractors are chosen and how they are monitored are the two most important aspects of dealing with subcontractors.
- Engineering managers and their subordinates more than likely will get involved in qualifying subcontractors, at least in the technical arena.
- Qualification of a subcontractor should be a very thorough process that examines the subcontractor's personnel and process control verification.
- The company's representative at a subcontractor's site should immediately report any unusual issues, such as a slowdown in the project or quality problems.
- Overseeing a subcontractor onsite is crucial, especially in foreign countries. For especially important projects, a reliable subordinate should be in place at the subcontractor's site on a full-time basis.
- In certain cases, it might become necessary to provide technical, personnel, or financial support to a subcontractor to help them complete the job.
- The engineering team should listen carefully to their subcontractors, who might have more expertise in a particular field.

REGULATORY RELATIONSHIPS

- Engineering managers and their subordinates will deal with numerous regulatory and certification agencies and companies during the life of a project.
- Engineering managers and their subordinates should deal with an expert agency in the particular country for which they are seeking certification.
- Engineering managers and their subordinates must make sure that a regulatory agency's certificate is complete and signed appropriately.
- When product failures occur during certification tests, the certification inspectors should be sent home. It is better to fix the problem while certification inspectors are not looking over your shoulder and to reschedule the tests after successful retesting.
- It is helpful to become familiar with the quirks of various certification inspectors encountered through the years.

Index